MOONSHOTS

MOONSHOTS

50 YEARS OF **NASA** SPACE EXPLORATION
SEEN THROUGH **HASSELBLAD** CAMERAS

Piers Bizony

Brimming with creative inspiration, how-to projects, and useful information to enrich your everyday life, Quarto Knows is a favorite destination for those pursuing their interests and passions. Visit our site and dig deeper with our books into your area of interest: Quarto Creates, Quarto Cooks, Quarto Homes, Quarto Lives, Quarto Drives, Quarto Explores, Quarto Gifts, or Quarto Kids.

© 2019 **Quarto Publishing Group USA Inc.**
Text © 2017 **Piers Bizony**

Paperback edition published in 2019. First publishing in 2017 by Voyaguer Press, an imprint of The Quarto Group, 100 Cummings Center, Suite 265D Beverly, MA 01915 USA
T (978) 282-9590 F (978) 283-2742 www.QuartoKnows.com

All rights reserved. No part of this book may be reproduced in any form without written permission of the copyright owners. All images in this book have been reproduced with the knowledge and prior consent of the artists concerned, and no responsibility is accepted by producer, publisher, or printer for any infringement of copyright or otherwise, arising from the contents of this publication. Every effort has been made to ensure that credits accurately comply with information supplied. We apologize for any inaccuracies that may have occurred and will resolve inaccurate or missing information in a subsequent reprinting of the book.

Motorbooks titles are available at discount for retail, wholesale, promotional, and bulk purchase. For details, contact the Special Sales Manager by email at specialsales@quarto.com or by mail at The Quarto Group, Attn: Special Sales Manager, 100 Cummings Center, Suite 265D Beverly, MA 01915 USA

10 9 8 7 6 5 4 3 2 1

ISBN: 978-0-7603-6676-9

The Library of Congress has cataloged the previous edition as follows:

Names: Bizony, Piers, editor, writer of supplemtary textual content.
Title: Moonshots : 50 years of NASA space exploration seen through Hasselblad
   cameras / [annotated] by Piers Bizony.
Description: Minneapolis, Minnesota : Quarto Publishing Group USA Inc.,
   Voyageur Press, 2017. | Includes index.
Identifiers: LCCN 2017017747 | ISBN 9780760352625 (hardbound)
Subjects: LCSH: Astronautics–United States–Pictorial works. |
   Astronautics–United States–History–Sources. | Space photography–United
   States. | United States. National Aeronautics and Space
   Administration–History–Sources.
Classification: LCC TL793.5 .M66 2017 | DDC 629.45/40973–dc23
LC record available at https://lccn.loc.gov/2017017747

Acquiring Editor: **Dennis Pernu**
Art Director: **James Kegley**
Layout: **Peter Hodkinson**, Spaced Design

Printed in China

"WE CHOOSE TO GO TO THE MOON IN THIS DECADE AND DO THE OTHER THINGS, NOT BECAUSE THEY ARE EASY, BUT BECAUSE THEY ARE HARD, BECAUSE THAT GOAL WILL SERVE TO ORGANIZE AND MEASURE THE BEST OF OUR ENERGIES AND SKILLS . . .
BUT IF I WERE TO SAY, MY FELLOW CITIZENS, THAT WE SHALL SEND TO THE MOON, 240,000 MILES AWAY FROM THE CONTROL STATION IN HOUSTON, A GIANT ROCKET MORE THAN 300 FEET TALL, THE LENGTH OF THIS FOOTBALL FIELD, MADE OF NEW METAL ALLOYS, SOME OF WHICH HAVE NOT YET BEEN INVENTED, CAPABLE OF STANDING HEAT AND STRESSES SEVERAL TIMES MORE THAN HAVE EVER BEEN EXPERIENCED, FITTED TOGETHER WITH A PRECISION BETTER THAN THE FINEST WATCH, CARRYING ALL THE EQUIPMENT NEEDED FOR PROPULSION, GUIDANCE, CONTROL, COMMUNICATIONS, FOOD, AND SURVIVAL, ON AN UNTRIED MISSION, TO AN UNKNOWN CELESTIAL BODY, AND THEN RETURN IT SAFELY TO EARTH, RE-ENTERING THE ATMOSPHERE AT SPEEDS OF OVER 25,000 MILES PER HOUR, CAUSING HEAT ABOUT HALF THAT OF THE TEMPERATURE OF THE SUN—ALMOST AS HOT AS IT IS HERE TODAY—THEN WE MUST BE BOLD."

President John F. Kennedy, September 12, 1962

Introduction 8

1 **THIS NEW OCEAN**

ORIGINS OF THE COLD WAR SPACE RACE   14

2 **APOLLO ASCENDING**

A TRIAL BY FIRE FOR HUMANS AND MACHINES   46

3 **A MAN ON THE MOON**

THE FIRST LUNAR TOUCHDOWN MISSIONS   84

4 MAKING TRACKS

ROVING THE MOON'S SURFACE ON WHEELS **144**

5 LIVING IN ORBIT

PERMANENTLY INHABITING THE ORBITAL REALM **196**

Index 238

Credits 240

INTRODUCTION

The photographs taken on film by American astronauts five decades ago still have the power to amaze us, in our digital age. They are a vivid reminder of past achievements, but also lay down a challenge for the future.

In this age of infinite information, what could be more familiar to us than those famous photos of men walking on the moon? At the same time, what could be less familiar than the hazardous visit to an airless alien landscape a quarter of a million miles away, conducted by just twelve people in all the world? It is not just distance but time that separates us from the achievements of this special dozen. Their bootprints, undisturbed to this day in the lunar soil, were tramped down half a century ago, before most of the seven billion people now alive were born; and of those twelve explorers, who were not precisely young when they reached the moon, only seven (as of early 2017) are still with us to recount their stories firsthand.

Those men, who were in their prime when the world knew each and every one of their names, and who are teasingly, cruelly, still full of youthful vigor in those photos . . . They are old now, and some of their names no longer come so easily to the forefront of popular recollection. Time is doing what it always does: separating us from the tangibility of events, turning the vivid exaltations and terrors of one generation into the vague memories and clinical schoolroom subjects of the next, and reducing the historic achievements of the many people who worked in or around Project Apollo to just two names: Neil Armstrong and Buzz Aldrin, the astronauts who made the first lunar landing in July 1969. This book stands as a tribute to them, their fellow astronauts and NASA colleagues, and the four hundred thousand men and women from all walks of life who contributed to the American space program during the busy decade before Apollo finally reached the moon. *Moonshots* is dedicated also to the people who persevered in the space adventure, up to (and beyond) the moment when the last of the operational winged space shuttle orbiters rolled to a halt at the end of the runway in the early morning of July 21, 2011.

It is normal for us to forget. If asked to identify the men who first conquered the Atlantic on wings, most people would recall Charles Lindbergh, who made his famous solo flight in 1927, but it was a British pair, John Alcock and Arthur Brown, who were the first aviators to cross that daunting ocean in June 1919, by a neat coincidence almost exactly fifty years before Apollo 11's landing craft touched down on the moon. They flew a modified World War I Vickers Vimy bomber with open cockpits, constructed from wood and canvas, held together by thin wires, and powered by a couple of sputtering Hispano-Suiza gasoline engines. At the time, this creaking biplane represented the pinnacle of technology, just as Apollo's similarly thin-skinned and fragile lunar module showcased the very best that 1960s ingenuity could devise.

Today the line for the restroom is on our minds as we hurtle across that same vast expanse of sea in air-conditioned jetliners flying ten times faster, and twenty times higher, than that old Vimy. We are headed toward important business meetings, family reunions, and wonderful vacations. We tend not to think of our own flights as epic. They are simply the fastest, most convenient, and least expensive means to our ends. Perhaps we pass the few hours that our crossings take by watching movies about men and women from long ago who did think in epic terms, and who were willing to risk their lives in terrifyingly fragile machines to pioneer a future that so many millions of us now take for granted.

So it is with the voyages of the first American astronauts. All of them believed that the hazards of rocket flight were worth taking for the benefit of future generations. The immediacy of their drama, and of the world's reaction to it, has faded over the last half a century because we remain unsure what kind of a future those astronauts were pioneering. Are we destined to go back to the moon? Are we supposed to set up camp on Mars? Such a goal may lie beyond this generation's capabilities, no matter what the Red Planet's fans try to tell us. We could get to Mars, but surviving there (and coming home again afterward) is another story. Are we supposed to hunt down asteroids? Not according to those who say that robotic probes are better suited to gather simple rock samples. Is the orbital realm where our spacefaring energies should be directed? We don't know any of the answers right now. Nevertheless, instinct impels us to believe that human space flight should continue, somehow, some way, for reasons as yet unclear.

All we do know is that if humans stopped venturing into space, we would mourn the loss of that capability and feel ourselves lessened as a species. Cash is not the problem. There is always enough money in the world to accomplish pretty much anything we want to, and still keep everyone fed. We could solve all the technical problems about life support and radiation too. The apparent holdup in our bid to become a spacefaring species is political, not mechanical.

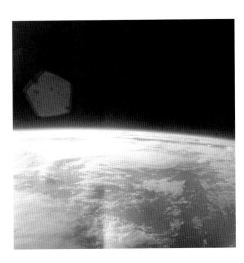

Astronaut Wally Schirra tries out a Hasselblad camera, and (at left) one of the views of the earth that he captured during his orbital Mercury mission in October 1962.

It is not (and never has been) NASA's job to tell us where we should go next in space. The role of this government agency is to carry out national space policy, not to determine it. Essentially, that choice belongs to the American people, and to their international allies. It belongs, dear reader, to you.

NASA's Photographic Legacy

To see our way forward, it always helps to have a clear view of the past. To this end, space enthusiasts hunt down rare and faded artifacts: NASA press releases with foxed corners, their paper as stiff and dry as leaves in the fall, or Apollo-themed magazines unearthed in thrift shops, usually with torn covers and crumpled pages. Even the copies preserved in libraries lack the freshness they would have had when first printed, because paper and ink change over time. Yes, time is doing what it always does: separating us from the vividness and intensity of events. And yet, in the vaults of NASA's Johnson Space Flight Center in Houston, Texas, a precious set of resources is held in cold storage in an effort to hold off time's depredations. These are the original film strips extracted from the cameras carried by the astronauts. The strips are not changeless, because nothing is permanent, but they degrade slowly because of the conditions in which they are stored.

What does constantly change is how those film frames are interpreted in other media. The first iterations were released hot in the wake of the astronauts' fiery homecomings. For instance, on August 11, 1969, just two weeks after Neil Armstrong, Buzz Aldrin, and Michael Collins had splashed down, *Life* magazine released a special edition featuring large-format color images from the Apollo 11 mission. The December 1969 issue of *National Geographic* had extensive color coverage. By that point, NASA's press office had sent out thousands of transparencies, prints, and other materials.

So what exactly were we looking at—those of us old enough to have seen those 1960s magazines after they first rolled off the presses? For guidance on this question, we have the wisdom of Ansel Adams (1902–1984), America's greatest landscape photographer, and one of the best photographic craftsmen who ever lived. What came out of his camera was not the final result. "The negative is comparable to the composer's score, and the print to its performance," he asserted. "Each performance differs in subtle ways." It was his total mastery of deriving positive paper prints in the darkroom that established his reputation. It helped that the negatives he pulled from his camera were similarly superb, but sometimes he would explore different ways of interpreting the same negative, adjusting tonal range from soft to high contrast, or perhaps using a more silver-rich paper stock, and so on.

In that same spirit, NASA did a great job creating initial "performances" from their original film frames. Then it was up to the countless magazines, books, and other printed versions to add their own touches, and here is when the serious diversions from what we might think of as the original images really began to kick in. After the editorial selections were made, and the layouts decided (and often, after square images had been cropped to suit rectangular page layouts), then the printing presses put in their ten cents' worth of interference. The choice of paper had an influence on how the ink settled, as did the mood swings of the presses. It was easy for the black to go down a little hazy, or the cyan to come down too hard. By the time the public saw any of NASA's famous images from space, they assuredly were looking at "performances" based on a wide range of variables, some controllable, others down to chance.

On the Christmas Eve of 1968, Apollo 8 astronaut Jim Lovell reported from lunar orbit, "The moon is essentially gray, no color. It looks like plaster of Paris or sort of a grayish beach sand." How was any magazine editor supposed to check the color balance of a gray moon, when even the best print systems of that era had their own biases? Gray is a difficult thing to render in pictures built from cyan, magenta, yellow, and black inks. Furthermore, even the best color film stocks balked at registering colorless grays. Photographers working with film knew this very well and selected their stock for the well-known biases, some great for architecture and landscape, others suited to portraiture, and so on. There

> **"THE FLIGHT EXPERIENCE ITSELF IS INCREDIBLE. IT'S ADDICTIVE. IT'S TRANSCENDENT. IT IS A VIEW OF THE GRAND PLAN OF ALL THINGS THAT IS SIMPLY UNFORGETTABLE."**
>
> **Mercury-Atlas 7 astronaut Scott Carpenter, June 1962**

never was a strip of color film that produced an absolutely "true" color rendering of a scene (whatever "true" means).

On top of that little problem, there just wasn't such a huge range of shots for picky editors to select from out of the Apollo haul, and that's why, over the years, we have tended to see the same ones reproduced time and again. A wonderful myth has gotten hold of some conspiracy enthusiasts. How could those lunar photos be real when they are so suspiciously perfect, as if created in a studio? If the doubters took the time to look through more than eight thousand frames retrieved from the handheld cameras of Apollo, and if they would only put themselves in the position of an art director trying to design a salable print product, they would struggle to select more than a few hundred that would pass muster. The sun's glare too often spoiled a shot. The spacesuits were so white and reflective they tended to bleach out, and so on. The astronauts were swamped by NASA's incessant demand for science tasks and had to point and shoot their cameras swiftly, without help from a viewfinder, while getting on with other work. It's amazing that they brought back as many good photos as they did, and much of the credit belongs to the cameras they wielded.

Wally Schirra, one of NASA's first team of astronauts, bought a Hasselblad roll film camera manufactured in Sweden and took it with him on his orbital flight aboard a one-man Mercury capsule in October 1962. His shots of Earth, hastily snapped through his tiny window, convinced NASA that Schirra had discovered a powerful tool, not just for space science but for public relations too. The film frames were large (two and a quarter inches square), yet the camera was surprisingly compact and mechanically rugged. The Carl Zeiss lenses were formidably sharp. Best of all, the Hasselblad was easy to reload. A conventional camera had to be opened up like a heart surgery patient, and its film carefully extracted and replaced. Astronauts could never have done this while wearing their spacesuit gloves. Hasselblad featured self-contained, instantly switchable film magazines. It was almost as if they had been designed with space explorers in mind.

Into the Digital Era

A quarter of a century after the first lunar touchdown, the technologies of film were under threat from a new and fast-growing interloper: digital imaging. In the mid-1990s, Michael Light, another American cameraman with an eye for large format precision, was struck by the similarities between some of the desert landscapes of his homeland, and the surface of the moon. He experimented with digital scanning of NASA film materials at a time when this was not so easy as it is now. He wondered if he might gain access to the very first generation of copies made from the precious original Apollo film strips, including many from the suite of automatic mapping cameras carried in the service modules of Apollos 15, 16, and 17. Sensing that a fresh new "performance" was justified, Light took on the daunting task of digitizing some of the materials and readying them for a large-format book, *Full Moon*, first published to worldwide acclaim in 1999.

Light's book certainly featured Apollo hardware and astronaut activities, but mainly he wanted to steer clear of familiar "greatest hits" materials and tell what was, at that time, a less well-known story, where the lunar terrain itself featured as his leading subject. *Full Moon* was a significant achievement, and *Moonshots* aims to serve as a companion piece by focusing precisely on those more famous images that everyone thinks they have already seen.

In the two decades since *Full Moon* first appeared, the relationship between film archives and digital interpretation has continued to evolve, and we think that yet another "performance" of historic NASA images is justified. We also believe that printing the results on paper is important. Long after the world's fleeting electronic archives have been lost or deleted, surviving copies of this book will stand (with other similarly respectful volumes) as a testimonial to America's first decades in space. Stored carefully, these pages should stay almost as pristine as the film frames in NASA's vaults. One of the reasons why *Moonshots* comes with a slipcase is to help prevent the fading that often happens when a book is shelved in daylight. This book commemorates (and it seems incredible to say this) the fiftieth anniversary of the first human voyages to the moon, and its publication coincides also with a period of stasis between the shutdown of the space shuttle program and the full emergence of NASA's new crewed launch technologies.

Many of us hope that the Apollo missions will not turn out to be our only experience of walking on another world, and that future space explorers will get the chance to continue where those first explorers left off.

What do you think?

> "IN 1969 I PREDICTED THAT THE SATURN V MOON ROCKET WOULD DOMINATE THE 1970s LIKE A COLOSSUS. IN THE EVENT, FAR FROM DOMINATING THE '70S, THE SATURN WAS DOMINATED BY THEM. BUT THE TIME WILL COME WHEN PROJECT APOLLO IS THE ONLY THING BY WHICH WE WILL REMEMBER THE UNITED STATES, OR INDEED THE WORLD OF OUR ANCESTORS, THE DISTANT PLANET EARTH."
>
> **Arthur C. Clarke, 1995**

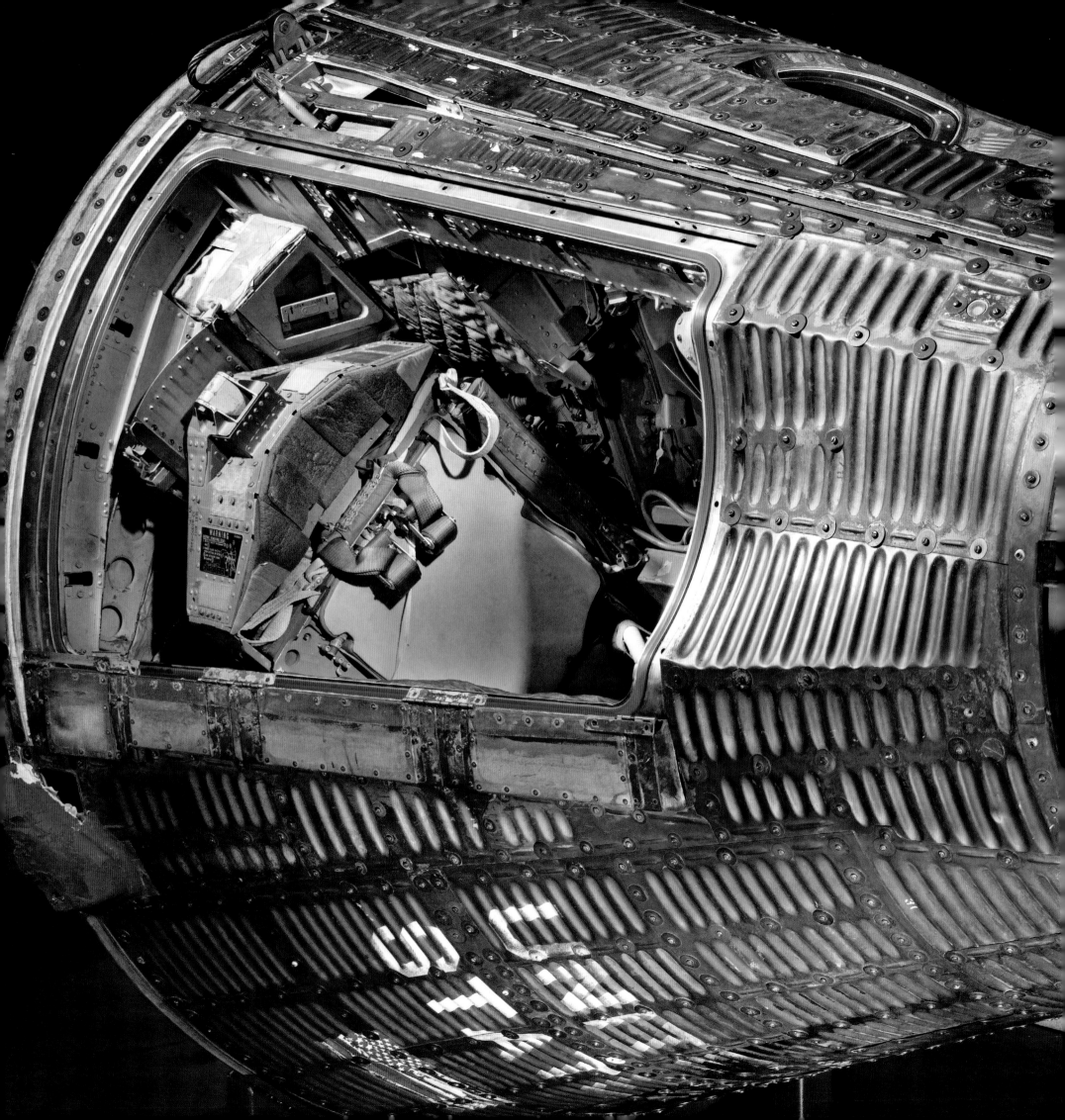

1 THIS NEW OCEAN

Our pioneering journeys into space were not propelled by Cold War rivalry alone. Since the dawn of time we had dreamed of flying into the cosmos, and in the mid-twentieth century, we had the tools that could make those dreams come true.

> **FIRST, INEVITABLY, COMES THE IDEA, THE FANTASY, THE FAIRY TALE. THEN, SCIENTIFIC CALCULATION. ULTIMATELY, FULFILLMENT CROWNS THE DREAM.**
>
> **Russian space theorist Konstantin Tsiolkovsky, 1926**

The story of space exploration is one of constant evolution. In the early twentieth century, rockets and space vehicles were the speculative visions of brilliant creative minds, such as Konstantin Tsiolkovsky, Robert Goddard, Hermann Oberth, and later, Wernher von Braun. Early science fiction writers such as Jules Verne and H. G. Wells inspired these forward-thinking pioneers. In the late 1950s, space technology emerged from the realm of pure imagination, and it became a major industrial, political, and military business. The world's first artificial satellite, Sputnik, was launched into orbit on October 4, 1957, using a converted ballistic missile. Essentially overnight, an epic rocket competition was unleashed between the United States and the USSR, the two Cold War superpowers. The world was captivated by the pioneering flights of satellites and probes, and then astronauts and cosmonauts, but the glamour of their missions could not entirely mask the fact that space exploration's opening chapters were dominated by political rivalry rather than science and peaceful exploration.

By the spring of 1961, the CIA and other American intelligence agencies had been warning for some time that the Soviets were preparing to launch a manned spacecraft. The rumors rapidly solidified into fact. On April 11, 1961, newly elected Democratic president John F. Kennedy appeared on an NBC early evening television program sponsored by Crest toothpaste. He and his wife, Jacqueline, talked with reporters Sander Vanocur and Ray Scherer about the difficulties of raising their small children and about the president's personal working style. Kennedy happened to remark that political events often appeared more complicated when viewed from inside the Oval Office than they did to the outside world. Even as he smiled for the cameras, he knew that a serious embarrassment awaited him in just a few hours' time. There was little he could do except brace himself.

At 1:07 a.m. Eastern Standard Time, NATO radar stations recorded the launch of a Soviet R-7 rocket, and fifteen minutes later a radio monitoring post in the Aleutian Islands detected unmistakable signs of live dialog with its human occupant. At 5:30 a.m. Washington time, the Moscow News radio channel announced the latest Soviet triumph in space. An alert journalist called NASA's launch center at Cape Canaveral in Florida to ask, could America catch up? Press officer John "Shorty" Powers was trying to grab a few hours' rest in his cramped office cot. He and many other staffers were working sixteen-hour days in the run-up to the first flight of NASA's one-man Mercury spacecraft. When the phone at his side rang in the predawn silence, he was irritable and unprepared. "Hey, what is this!" he yelled into the phone. "We're all asleep down here!" Later that morning the headlines read: SOVIETS PUT MAN IN SPACE. SPOKESMAN SAYS U.S. ASLEEP.

Yuri Alekseyevich Gagarin, a twenty-seven-year-old Soviet Air Force pilot, had just become the first human ever to fly in space, aboard a small capsule named Vostok ("East"). Kennedy held an uncomfortable press conference. Normally a self-confident and eloquent public performer, he seemed distinctly less sure of himself than usual. He was asked, "Mr. President, a member of Congress today said he was tired of seeing the United States coming second to Russia in the space field. What is the prospect that we will catch up?" He replied, "However tired anybody may be—and no one is more tired than I am—it is going to take some time. The news will be worse before it gets better. We are, I hope, going to go into other areas where we can be first, and which will bring perhaps more long-range benefits to mankind. But we are behind."

Gagarin's short journey into space was one of the most important events of the twentieth century—not just for Russia, but for America too, where an industrial shake up of colossal proportions was unleashed in response. Much of the fabric of modern technology was designed with the Space Race (and a potential nuclear missile war) in mind. Microchips were developed because 1950s circuitry was too big, too heavy, and too delicate to fit inside rockets and missiles. The internet emerged from an attack-proof communications network laid down by the Advanced Research Projects Agency (ARPA), a pre-NASA government department that planned for America's future in space, among other things. The global communications industry developed with incredible speed after the invention of satellites. In all likelihood, these technologies would have come along anyway, but probably not as fast as they did. And all because Yuri Gagarin, a good-natured young jet pilot and now the world's first space traveler, had thrown down a challenge to the most powerful nation on earth.

Dr. John Logsdon, former head of the Space Policy Institute in Washington, D.C., has advised a succession of presidents and explains the impact of Gagarin's flight on the American psyche: "It was a sudden rebalancing of our power relationship with the Soviet Union, because of the clear demonstration that, if they wanted to, they could send a nuclear warhead across intercontinental distances, right into the heart of Fortress America. There was an uproar. How did we get beaten by this supposedly backward country?"

Responding to the Challenge

Kennedy was agitated at the global response to Gagarin's flight. He paced his office at the White House, asking his advisors, "What can we do? How can we catch up?" His science advisor Jerome Wiesner opposed any reckless rush, but the president wanted an urgent response. "If somebody can just tell me how to catch up. Let's find somebody. Anybody. I don't care if it's the janitor over there, if he knows how," the president said. He deliberately made these remarks within earshot of Hugh Sidey, a senior journalist from *Life* magazine. All of a sudden the president wanted to be seen as an advocate for space.

Three days later, he suffered another more serious defeat. A 1,300-strong force of exiled Cubans supported by the CIA landed at the Bay of Pigs in Cuba with the aim of destroying Fidel Castro's communist regime. Kennedy had approved the scheme, but Castro's troops learned of the operation ahead of time and were waiting on the beaches. The raid was a disaster. The Kennedy administration seemed to be faltering in its first one hundred days, the traditional honeymoon period during which a new president was supposed to stamp his particular vision on the country. He immediately turned to space as a means of reviving his political credibility. In a pivotal memo of April 20, he asked his vice president, Lyndon Johnson: "Do we have a chance of beating the Soviets by putting a laboratory in space, or by a trip around the moon, or by a rocket to land on the moon, or by a rocket to go to the moon and back with a man? Is there any other space program which promises dramatic results in which we could win?"

Johnson, a keen supporter of space technology, convened a panel of experts to help decide the matter, but he already knew which option he favored. Wiesner commented afterward, "Johnson went around the room saying, 'We've got a terribly important decision to make. Shall we put a man on the moon?' And everybody said, 'Yes.' And he said, 'Thank you,' and reported to the president that the panel said we should put a man on the moon." Or to put it another way, in the words of Pulitzer Prize–winning historian Walter McDougall, "Johnson sent the president a report so loaded with assumptions that a moon landing was the inescapable conclusion."

As a powerful Democratic senator supporting the Democratic cause throughout the Eisenhower presidency, Johnson had overseen the creation of NASA in 1958, in response to Sputnik. Now he steered it toward the moon because he believed in the value of that goal, not just as a sturdy reply to Soviet domination, but also as a genuine expression of American destiny. His politics had been forged in the 1930s Roosevelt New Deal era. Kennedy was somewhat more pragmatic and went along with the lunar landing idea because it was expedient for him at the time, but when he did decide to push for it, he did so by delivering some of the most eloquent and memorable speeches in all of political history.

A final decision hinged on NASA getting its manned space program off the ground. On May 5, just twenty-three days after Gagarin had flown, astronaut Alan Shepard was launched atop a small Redstone missile. His flight wasn't an orbit, merely a ballistic hop of fifteen minutes' duration. In contrast to Vostok's orbital velocity of 17,500 miles per hour, Shepard's Mercury achieved only 5,000 miles per hour. Vostok girdled the globe while the Mercury splashed down in the Atlantic just 320 miles from its launch site. But this cannonball flight was enough to prove NASA's basic capabilities. Stunned by the potential costs, Kennedy nevertheless decided to support NASA's Apollo lunar project. In a historic speech before Congress on May 25, 1961, he said, "I believe that this nation should commit itself to achieving the goal, before this decade is out, of landing a man on the moon, and returning him safely to the Earth. No single space project in this period will be more impressive to mankind, or more important for the long-range exploration of space, and none will be so difficult or expensive to accomplish."

Not so many months before, Kennedy and his administration had struggled to find anyone willing to lead NASA, because it had seemed, at first, like a small agency engaged on a high-risk, underfunded Buck Rogers experiment. Seventeen highly qualified people turned down the job. Fortunately, an acquaintance of Lyndon Johnson was willing to take on the task. James Edwin Webb took up Kennedy's challenge, and rapidly gathered together the political, financial, real estate, and technological resources that NASA would need. A master of political machination, Webb took care to spread the NASA activities far beyond just the Florida launch site. That choice of seaboard location was dictated by technical considerations to do with the rotation of the Earth, and by the need to keep rockets away from built-up areas. No such justification existed for building NASA's new mission control center far away in a dusty patch of Texas scrubland, but Webb knew that it made excellent political sense. His skillful stewardship gave NASA the momentum it needed.

Throughout the mid-1960s, NASA demonstrated to the world a "can-do" spirit with its two-man Gemini spacecraft, rehearsing most of the techniques we now take for granted in space travel, such as space walks, orbital rendezvous, and dockings. However, on January 27, 1967, the crew of the first Apollo spacecraft, Ed White, Roger Chaffee, and Virgil "Gus"

> " I TRAINED HARD FOR MANY YEARS TO FLY AROUND THE MOON AND TAKE A CLOSE LOOK AT IT, BECAUSE I THOUGHT THAT'S WHAT THE MISSION WAS. WHEN WE GOT HOME, I REALIZED WE HAD DISCOVERED SOMETHING MUCH MORE PRECIOUS OUT THERE: THE EARTH. I BELIEVE THE ENVIRONMENTAL MOVEMENT WAS TREMENDOUSLY INSPIRED BY THOSE MISSIONS. THAT ALONE IS WORTH THE RELATIVELY FEW BILLIONS OF DOLLARS THAT WE SPENT ON APOLLO. "
>
> **Apollo 8 astronaut Bill Anders**

Grissom, died when their capsule caught fire just sitting on the launch pad during what should have been a routine communications test. The subsequent investigation revealed faults in the Apollo's design, and management errors both within NASA and among the manufacturing contractors were generated in part by everyone's sense of urgency as they rushed to meet Kennedy's famous deadline for reaching the moon.

Speeding Up the Race

Apollo 7, the first successful crewed flight in the series, was launched into Earth orbit in October 1968, atop a half-sized variant of the Saturn rocket known as the 1B. Mission commander Wally Schirra, and colleagues Donn Eisele and Walter Cunningham, proved that the spacecraft was ready for action. The huge Saturn V moon rocket had failed badly during an unmanned test flight, so it took some nerve for NASA to put astronauts on the next available one and send them straight to the moon aboard Apollo 8, regardless of the fact that they could not land, because the landing module was also experiencing development problems.

Subsequent Apollo missions were launched in quick succession, one flight following another at a pace that seems incredible in retrospect. Apollo 9 tested NASA's spidery lunar module (LM) in Earth orbit. The next mission swooped to within a tantalizing few miles of a lunar touchdown, and at last, on July 20, 1969, an ancient dream was fulfilled. Men had walked on the moon. Who could have guessed that in just another three years' time it would once again slip beyond our reach for at least another generation?

In his celebrated memoir, *Carrying the Fire*, Apollo 11's command module pilot Michael Collins recalled worrying about what he would say as he concentrated on the switch settings to blast the ship out of Earth orbit and into deep space. "If we're going to leave the gravitational field of Earth, what are we going to say? We should invoke Christopher Columbus, or a primordial reptile coming up out of the swamps onto dry land for the first time, or go back through the sweep of history and say something meaningful.

> **IF WE CHOOSE TO TRAVEL INTO SPACE IN THE FUTURE, IT SHOULD NOT JUST BE TO CONQUER NEW WORLDS AND REPLICATE OUR SPRAWLING SHOPPING MALLS ON OTHER PLANETS. WE SHOULD GO OUT THERE TO DISCOVER MORE ABOUT OURSELVES, AND TO LOOK BACK, WITH WISER AND MORE APPRECIATIVE EYES, AT THE WORLD WE LEAVE BEHIND.**
>
> **Apollo 8 astronaut Bill Anders**

Instead, all we had was our technical jargon." If an astronaut was dismayed by his own terminology, just imagine its demoralizing effect on the rest of humanity. The famed novelist and journalist Norman Mailer visited NASA's Florida launch complex in that epochal summer of 1969 and saw a technocratic army of white middle-class men sitting at their control panels in neat, obedient rows, most of them dressed alike in white shirts and black ties. Listening to them as they readied Apollo 11 for launch, he thought that "their strength apparently derived from being cogs in a machine. They spoke in a language not fit for a computer of events that might yet dislocate eternity." Neither NASA nor its astronauts were trained for the ambiguities of art, literature, or poetry. The problem turned out to be that poetry was exactly what the rest of the world needed if it was to appreciate Apollo's achievements properly. As Michael Collins remarked after his historic mission, "A future flight should include a poet, a priest, and a philosopher. Then we might get a much better idea of what we saw."

Fifty years later, we can perhaps appreciate—even more than the world did at the time—the incredible achievements of the Gemini and Apollo series. Even after the Apollo 13 scare, NASA's space machinery seemed highly advanced and futuristic for its time. From our perspective, it all looks insanely delicate and dangerous. We, who have been raised on personal tablets and smartphones, can hardly understand the idea of guidance computers with switches and dials, let alone lunar landing vehicles swathed in silver and gold foils that you could have punched a fist through without even scratching yourself. Apollo was the product of a different age, and its success has much to teach our arguably more pampered and risk averse society about stretching the limits of the possible and getting things done.

Anyway, that supposedly antique technology still had some surprises up its sleeve. In April 2007, visitors to a NASA museum exhibit discovered that the apparently lifeless artifacts on display from a long-vanished era of space exploration were not quite so dead after all. As if in some technological version of *Raiders of the Lost Ark*, a team of awestruck technicians was carefully removing access panels on a spare, unused Apollo spacecraft, revealing pristine mechanisms within.

This was backup hardware for the 1975 Apollo-Soyuz docking mission that signaled the end of the Cold War rivalry in space. All the systems were intact. The investigating team was granted permission to strip down the one small section of the spacecraft that most interested them. They wanted the answer to a problem that had beaten the best minds in modern space engineering. At the end of its mission, and just prior to reentry into the earth's atmosphere, an Apollo capsule had to separate cleanly from the cylindrical service module that had carried most of its air, water, electrical supplies, and propulsion fuels. Many dozens of power wires, fluid cables, and data conduits had to be disconnected in an instant. How was this accomplished reliably, and in a split second, by those cranky old engineers from a bygone age?

The team discovered inside the Apollo a tiny set of explosive guillotines, through which all the cables and lines snaked in a neat bundle. When the time came for the capsule to drop away, the guillotines sliced through metal and plastic as though they were butter. A backup set of guillotines, powered by an entirely separate electrical system, insured against failure. And all in a box the size of a car battery. It seems as if Apollo still has lessons for us, as NASA gears up for new voyages in the coming generation. The upcoming Orion spacecraft looks very much like Apollo's bigger brother.

Today we often hear that the on-board Apollo Guidance Computer (AGC) had less power than a modern digital watch. But it all depends on what we mean by "power." In its way, it was one of the most capable computers ever invented. It absorbed data from a complex gyroscopic inertial navigation system, allied to an optical star telescope and two radar range finders. It also mediated between the astronauts and the thrusters and rocket engines that drove their ship through space.

NASA favored ruggedness over sophistication. Even when Apollo 12 was struck by lightning soon after launch in November 1969, and the interior of the capsule blacked out for a moment, its AGC cycled back to life in less than a second. It was a reliable piece of equipment because it had to be. There is more to great technology than complication for its own sake, and this is another lesson that Apollo can teach us.

Humans or Machines?

Discussion about Apollo usually calls to mind the big question about space exploration. Do we need to send people? Robotic missions have already shown us that there are other worlds, far beyond Earth, that are potentially within our reach. Today, the International Space Station is just the first stage in a journey that may take us deeper into the solar system. When we see the canyons of Mars through the sharp digital eyes of robot rovers, who among us does not imagine standing among those same landscapes and seeing them with our own eyes? Fifty years ago, such dreams were science fiction. Then, just for a moment at least, they started coming true. If we are to continue our explorations of space, there will always need to be at least some human involvement, or else it will not be a fully human adventure.

Arthur C. Clarke believed that "one day we will not travel in spaceships. We will be spaceships." In a sense, we already are. Robot probes are extensions of our minds and our physical reach, allowing us to scrabble in the soils of distant worlds with remotely operated claws. Does that mean we no longer have to go to the trouble of turning up in person to clutch handfuls of alien dust in our own (albeit, gloved) hands? Logic suggests that machine probes are the safest and most cost-efficient tools for space exploration. Instinct and emotion cause many of us to think differently; otherwise no astronaut, cosmonaut, taikonaut, or private adventurer would ever leave the ground. Whatever the economic or political arguments may be, it has become almost impossible for us to imagine a world without people in space helmets. Should the day ever arrive when we are incapable of launching anyone into the limitless expanse beyond the thin skin of our atmosphere, surely this will indicate that something has gone dreadfully wrong back on Earth.

About the Cameras

Wherever humans go, they want to record what they see and share their experiences with others. Scientists often need complex data from space missions—temperatures, radiation counts, spectral analyses, mineral ratios, and so on—but almost all the popular awareness of human space exploration comes from the more straightforward visual record. While NASA's first team of seven astronauts was preparing for their Mercury missions, relatively little thought was given to the use of normal film cameras. The priority was to get those men up and beat the Russians. An early design for the Mercury spacecraft even proposed leaving out the window, because it was a structural weakness.

The astronauts were self-motivated aviators, and they rebelled in no uncertain terms about that window. When John Glenn was launched to orbit in February 1962, he carried with him a small Ansco Autoset 35-millimeter camera, manufactured by the Japanese company Minolta. He barely had time to use it. Anyway, in the tense political circumstances at that time, NASA worried that taking snapshots of the Earth from space might come across as an attempt to acquire spy photographs while the astronauts were passing over other countries. Various extremely secret surveillance satellite programs were getting under way at the same time as project Mercury, with budgets and resources that almost matched NASA's, so there was a great deal of sensitivity about the use of cameras in general. It quickly became obvious that astronauts operating in the spotlight of national publicity should be allowed to carry cameras, and when Wally Schirra tried out his personal store-bought Hasselblad 550C during his orbital Mercury flight in October 1962, he fell in love with it, and the obvious scientific and publicity value of recording a space mission photographically was established. Today it seems bizarre that there ever could have been a time when cameras were an afterthought in the astronaut story.

On Apollo 8, Hasselblad EL cameras were used for the first time. They featured battery-powered motors that wound on the film and readied the shutter at the press of a button, greatly speeding up the picture-taking process. Each replaceable magazine could hold sufficient film for 160 color Ektachrome or 200 black and white Panatomic-X frames, captured on specially thin film bases developed for NASA by Kodak. From Apollo 11 onward, a further upgrade to the Hasselblad, the 500EL, was used. A transparent glass Réseau plate, engraved with grid markings, was installed between the film magazine and the camera body. From these markings, it was possible for photo analysts to calibrate the distances and heights of objects in the frames. Those thin black crosses are among the most iconic elements in any Apollo photo.

The Hasselblad Company today has, of course, gone digital. Its latest cameras can deliver images of almost unbelievable sharpness and resolution, but no matter what great new systems it unveils in the future, there is no doubt that this famous brand will always be celebrated for its pivotal role in the exploration of space.

HEAVENLY TWINS

The two-seater Gemini craft was NASA's sportiest, with gull-wing doors for spacewalking (or escape by ejection) and a complex array of thrusters. It was also the first space vehicle to carry an onboard computer to calculate rocket firings. Not that the computer was up to much by modern standards. Gemini astronaut John Young gave an unflattering description: "Imagine a box with a lot of coat hangers bent in all directions. It was basically wiring connecting thousands of individual transistors, with tiny electromagnets for memory nodes, very crude compared to today's microprocessors." Yet this basic computer, with the equivalent of less than 16 kilobytes of memory, enabled Gemini to change orbits and locate other vehicles for docking, the most essential of all space maneuvers. NASA's astronauts needed to practice docking ahead of the lunar missions, and Gemini was their teaching tool.

After two successful unmanned shots, Gemini III took off on March 23, 1965, with Virgil "Gus" Grissom and John Young at the controls for a five-hour orbital test of their "Gusmobile." As it turned out, the machines performed better than the humans on this first manned trip of the capsule. Mission Control did not yet trust Gemini's new computer. Grissom had been instructed not to use it for reentry if the onboard displays disagreed with NASA's earlier prediction of where the capsule would splash down. Sure enough, the computer disagreed, so Grissom switched it off—and Gemini III plunged into the Atlantic way off course, more than 60 miles from the nearest rescue ships. Young said afterward, "We knew a few days later, when NASA checked the figures, that if we'd gone with the computer, we could have parked our capsule right alongside the recovery ship."

In June 1965, Jim McDivitt's copilot Ed White conducted America's first space walk from Gemini IV. He floated outside the capsule for half an hour, connected only by a thin umbilical cord. "This is great! I don't want to come back inside!" he said. Overinflation of his suit made it hard for him to get back inside the Gemini, but he managed.

A Dangerous Docking

Neil Armstrong and Dave Scott blasted off aboard Gemini VIII on March 16, 1966, to link up with an unmanned Agena target vehicle launched ahead of them. Armstrong steered the ship to the first-ever docking. Moments later, he reported, "We've got serious problems. We're tumbling end over end!" Armstrong immediately backed the capsule out of the Agena, only to find that the tumbling became worse. A thruster was jammed in the "on" position and Gemini VIII began to spin violently out of control. Armstrong and Scott could barely keep conscious. Armstrong took a bold decision and cut power to all the docking and control thrusters,

The Gemini spacecraft were launched to orbit by converted Titan missiles originally designed to carry nuclear warheads. The Soviets, too, had adapted missiles into space launchers.

hoping to kill the rogue thruster in the process. It worked, but now he was left with just the small reentry control thrusters to steer with. He brought the ship back under control and found that he'd used up most of his fuel. He made an emergency splashdown with what he had left. That spectacular performance tipped him as the man most likely to keep cool in a potential moon mission.

The Gemini series was packed full of drama. For instance, Tom Stafford and Gene Cernan flew Gemini IX in June 1966, and when they came alongside their target vehicle, they found that its nose fairing had not separated properly, leaving the docking collar blocked. Stafford described the misshapen craft as an "angry alligator."

There were plenty of successes too. The Gemini ships practiced rendezvous and docking techniques that are essential for any mission involving more than one spacecraft. Gemini also helped solve the problems of spacewalking, which turned out to be much less exhausting and more productive after handholds and footrests were built into the space hardware at strategic intervals.

"WE SET SAIL ON THIS NEW SEA BECAUSE THERE IS NEW KNOWLEDGE TO BE GAINED, AND NEW RIGHTS TO BE WON, AND THEY MUST BE WON AND USED FOR THE PROGRESS OF ALL PEOPLE. FOR SPACE SCIENCE, LIKE NUCLEAR SCIENCE AND ALL TECHNOLOGY, HAS NO CONSCIENCE OF ITS OWN. WHETHER IT WILL BECOME A FORCE FOR GOOD OR ILL DEPENDS ON MAN, AND ONLY IF THE UNITED STATES OCCUPIES A POSITION OF PRE-EMINENCE CAN WE HELP DECIDE WHETHER THIS NEW OCEAN WILL BE A SEA OF PEACE OR A NEW TERRIFYING THEATER OF WAR."

President John F. Kennedy, September 12, 1962

An interior view of Gemini IV, from which Edward White made the historic first U.S. spacewalk on June 3, 1965. It is now on display at the Smithsonian National Air and Space Museum, Washington, D.C.

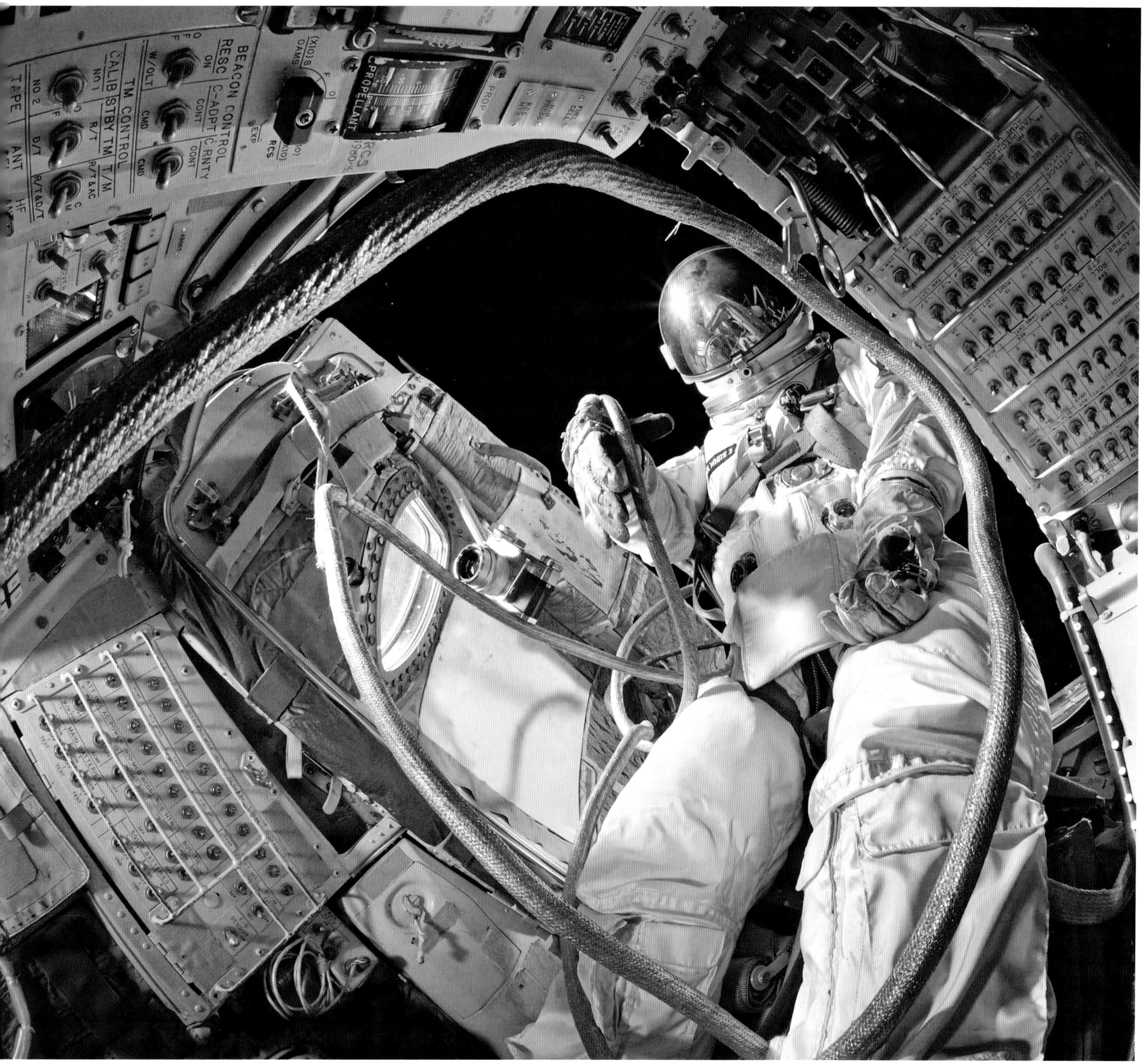
BEACON CONTROL
TM CONTROL
RCS
PROPELLANT

This time-lapse photo of a Gemini simulator approaching its mocked-up target demonstrates the project's chief ambition: to master the art of orbital rendezvous.

Clad in lint-free garments, astronauts Virgil Grissom (left) and John Young conduct a January 1965 communications test in their Gemini III spacecraft, just delivered to NASA's Kennedy Space Center in Florida.

Gemini IV copilot Ed White conducts the first American spacewalk on June 3, 1965. He found the experience exhilarating, although his suit ballooned in the vacuum of space, and he found it hard to get back into the spacecraft.

Gemini VI seen from Gemini VII as they make the first space rendezvous in December 1965. The Soviets claimed to have done something similar, but did not have the genuine capability at that time.

Gemini VII photographed through one of the hatch windows of Gemini VI during rendezvous and station keeping 160 miles above the earth. By now, the Hasselblad cameras had proved their value as high-resolution image-making tools for space flight.

A rear view of Gemini VII as the two spacecraft coast in their shared orbit, less than 40 feet apart, but both traveling at 17,500 miles per hour.

An adapted Atlas missile, first used as a launcher during the Mercury program lofts an unmanned Agena vehicle to serve as the docking target for a Gemini spacecraft.

> "IF WE DIE, WE WANT PEOPLE TO ACCEPT IT. WE'RE IN A RISKY BUSINESS."
>
> **Astronaut Virgil "Gus" Grissom, March 1965**

MAKING A RENDEZVOUS WITH ANOTHER CRAFT IN THE VASTNESS OF SPACE: WHICH NATION WAS THE FIRST TO TRY OUT THIS ESSENTIAL ORBITAL TECHNIQUE?

On August 11, 1962, two years before the first manned Gemini mission got under way, Soviet cosmonaut Andrian Nikolayev was launched in the same kind of Vostok capsule that had propelled Yuri Gagarin into the history books. Next day Pavel Popovich went up in another Vostok. For the first time, two people were in space simultaneously, in different ships. The launches were timed so that the second Vostok briefly came within a few miles of the first: a cosmic clay pigeon shoot that enabled the Kremlin spin doctors to claim a space rendezvous.

The two ships drifted apart and could not regain their close formation because they lacked maneuvering thrusters. However, appearances counted for a great deal back then. A number of space professionals in the West were fooled into thinking that the Soviets had developed genuine rendezvous skills.

In December 1965, Gemini VI astronauts Jim Stafford and Wally Schirra steered their ship until it came nose-to-nose Gemini VII. They could see Jim Lovell and Frank Borman grinning at them through the windows of the other ship. Afterward, Schirra said, "No way did the Russians rendezvous. That was just a passing glance, like seeing a pretty girl on the pavement from your car window before the flow of traffic whizzes you on. With Gemini, we could cut across the traffic and say hello. Now, that's what you call a rendezvous."

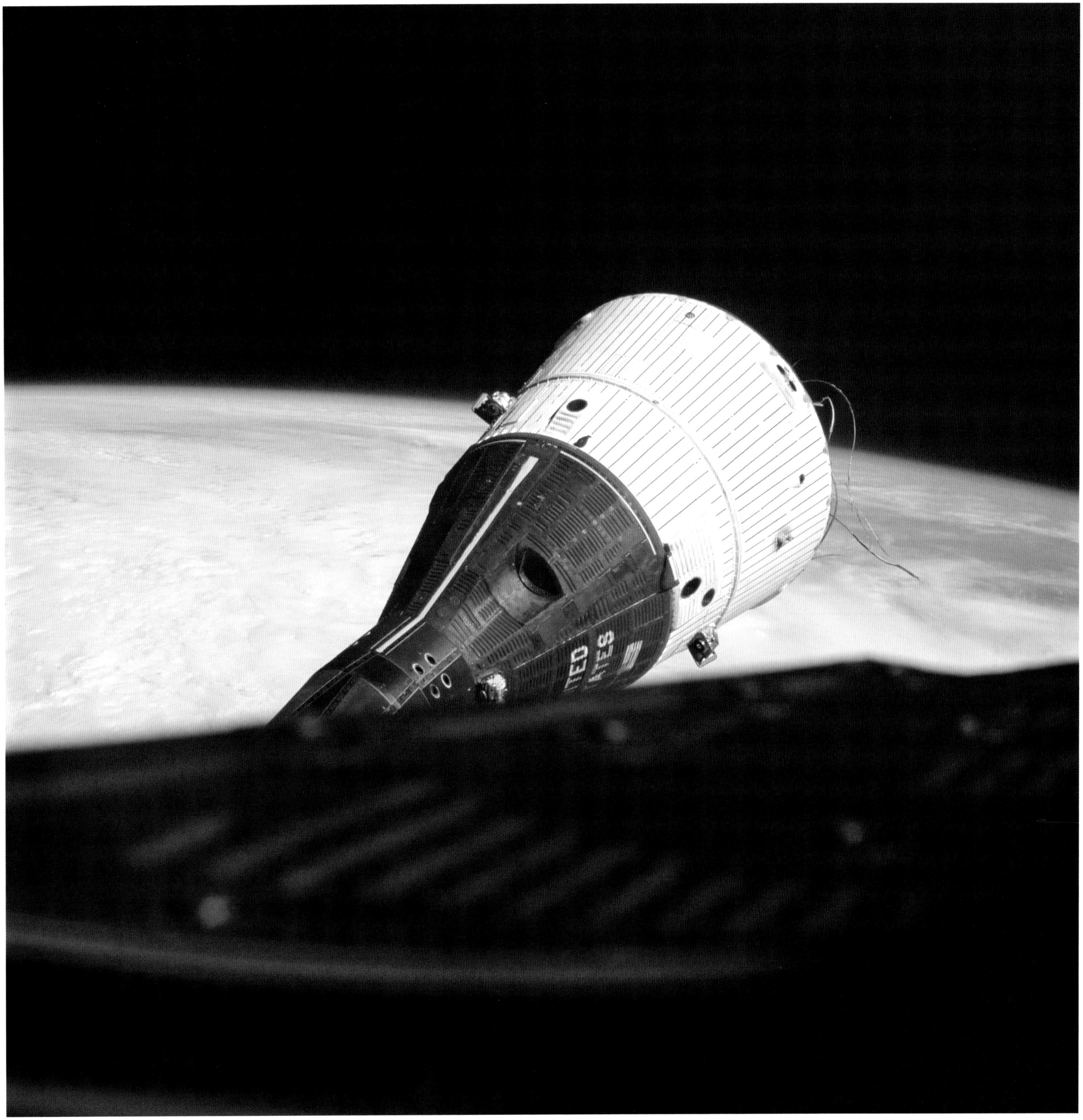

An Agena waits patiently for Neil Armstrong and David Scott to make the first docking on March 16, 1966. Seconds later, Gemini VIII got violently out of control.

Gemini VIII's commander Neil Armstrong handled a potentially terrifying situation with great calmness. As a pilot, he was seemingly unflappable, whether in the air or in space.

Gemini VIII's view of the Agena vehicle's docking collar, moments before a near-lethal thruster problem put the two astronauts into extreme jeopardy.

A lucky circumstance of lighting allows us to see Buzz Aldrin's face through his visor (the gold sunshield is raised) during his successful Gemini XII spacewalk.

Commander Tom Stafford photographed by colleague Gene Cernan aboard Gemini IX, which turned out to be a difficult and frustrating mission.

David Scott (at left) and Neil Armstrong look glad to be home after Gemini VIII's emergency return to Earth.

Stafford described Gemini IX-A's malfunctioning target vehicle as an "angry alligator." The nose shroud failed to separate, and a docking was impossible.

Pete Conrad clambers carefully out of Gemini XI. The green dye in the water is a location aide for the recovery helicopter.

Gene Cernan hastily grabbed a few photos of Gemini IX during his difficult and exhausting spacewalk. Future missions included handholds and footrests for purchase.

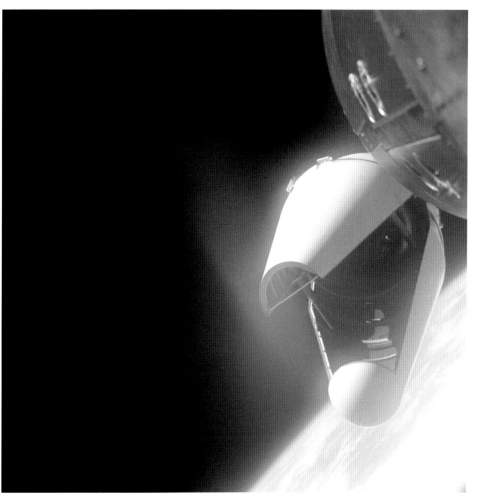

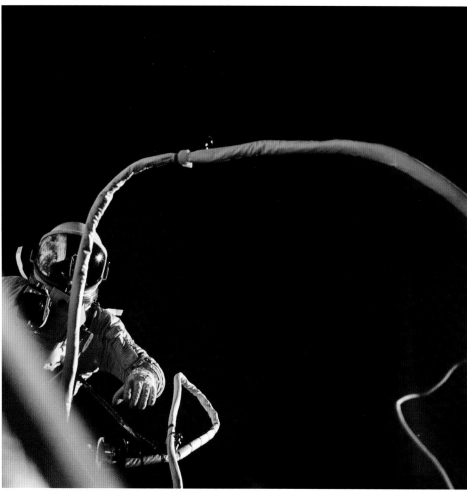

The Apollo 8 spacecraft mounted on its Saturn V launch vehicle stands on Pad A, Launch Complex 39, prior to carrying three men out of Earth orbit and into deep space for the first time in history.

2 APOLLO ASCENDING

After the success of Gemini, Apollo's bid for the moon began badly, with the deaths of three astronauts. Recovery of momentum was essential, but tough.

> **"WE ARE GOING TO HAVE FAILURES. THERE ARE GOING TO BE SACRIFICES MADE IN THE PROGRAM. WE'VE BEEN LUCKY SO FAR. WE HOPE THAT IF ANYTHING HAPPENS TO US IT WILL NOT DELAY THE PROGRAM. THE CONQUEST OF SPACE IS WORTH THE RISK OF LIFE."**
>
> **Astronaut Virgil "Gus" Grissom, March 1965**

In January 27, 1967, Virgil "Gus" Grissom and his crewmates, Ed White and Roger Chaffee, clambered into the first flight-ready Apollo, atop a half-sized version of the Saturn rocket, the 1B. This was a routine checkout procedure. They would do a simulated countdown with all internal systems running, but would not actually ignite the Saturn's engines for takeoff. There was no fuel in the tanks. Nevertheless, the ship exploded.

Chaffee complained that the spacecraft interior smelled of sour milk. Then the radio system glitched. Frustrated, Grissom shouted, "How the hell are we supposed to talk to mission control from space when we can't even reach them on the ground!" The mood was tense as the launch pad workers locked Apollo's heavy hatch into place, sealing the crew inside. Five hours into the test, Grissom's garbled voice on the crackling radio link said, "We've got a fire in the capsule." A few seconds later, another voice, possibly White's, was more urgent. "Hey, we're burning up in here!" There was a scream of pain. Then just a hiss of static as the radio went dead. Suddenly the side of the capsule split open. The top of the launch gantry was engulfed in thick smoke. The pad crew tried desperately to get the astronauts out. The fumes were impenetrable, and the heat overpowering. It took four minutes to open the Apollo's hatch. By then, the astronauts inside were dead.

Nothing in the first innocent decade of its existence had prepared NASA for this kind of darkness. Two prospective Gemini astronauts had been killed in a jet plane crash, but the aviation community was used to that kind of a bad day. This was different. Investigators digging into the circumstances of the Apollo1 disaster found that certain fire risks had been overlooked, such was the mood of confidence at that time after the many successes of the Gemini program. NASA had chosen to pump pure oxygen into the Apollo's cabin for the astronauts to breathe. It was pressurized at higher than normal levels so that the capsule would not be contaminated by dust or moisture from the Floridian air. Oxygen becomes more flammable under pressure, and somehow everyone had missed this danger. Some of the blame was also laid at the door of the manufacturing companies. Untidy wiring frayed against the sharp edge of a small access panel may have caused a short circuit spark.

James Webb, NASA's chief administrator, spent the next several months defending his agency's performance during a series of intense, and sometimes hostile, congressional and senatorial investigations. Nothing was out of bounds, from wiring bundles in the spacecraft, to NASA's management structures, or its business relationships with its contractors. At one point, a question about vending machines in an Apollo plant threatened to escalate into a corruption panic reaching into the heart of American government, but the stink faded almost as soon as it arose.

It is satisfying to run a big space project when things are going smoothly, and for sure, Webb and his people had ridden high throughout most of the 1960s. Now he faced his toughest challenge. Toward the end of the inquiry, he made an impassioned plea to his interrogators. "We have a responsibility to purge what is bad in the system we together have created and supported. We have an even greater responsibility to preserve what is good, and what represents still, at this hour, a high point in all mankind's vision," he said. Similarly eloquent testimony from Frank Borman, speaking on behalf of all the astronauts, also helped ease the tensions, while NASA and its contractors worked to solve the problems that had contributed to the fire.

We are not angels. The machines of Apollo were the incredible but not entirely perfect products of ordinary men and women working as best they could within a new and hugely complicated organization, setting about a dangerous task that no one had ever attempted before—and all in the public eye, with seemingly no detail spared from scrutiny. Yet they reached into the heavens despite their flaws, and in the face of incredible pressures: personal, professional, emotional. What gave Apollo its grandeur was not so much the awe-inspiring height of its rocket, nor the power of its engines, or the dazzling complexity of its innards, but its people.

The best way to regain public trust was to get flying again. On October 11, 1968, Apollo 7 reached Earth orbit with a crew aboard. Donn Eisele, Walter Cunningham, and commander Wally Schirra's task was to prove that their spacecraft functioned properly after all the redesigns.

Additionally, they fired the hydrogen-fueled upper rocket stage, the S-IVB, whose job, in future missions, would be to push the combined Apollo command module (CM) and spiderlike lunar module (LM) out of Earth orbit and toward the moon.

After a successful rocket burn putting Apollo 7 into a higher orbit, the command module disconnected from the S-IVB, turned around, and conducted a simulated docking approach, as if preparing to pull a lunar module out of the S-IVB's protective shroud. Finally, the large engine at the rear of Apollo 7's cylindrical service module was fired several times, making sure that it could brake future crews into lunar orbit and boost them away again for the trip back to Earth. The astronauts were prickly with mission control, all afflicted with head colds, and anxious to concentrate on their new and untried ship, but everything worked. Apollo's progress toward the moon was back on track.

Hazards in Perspective

A similarly terrible disaster in the Soviet space program helped put the Apollo fire into context. On April 23, 1967, an R-7 rocket carrying the first of Russia's new Soyuz spacecraft was propped up against the gantry at Baikonur, ready for launch. Archive footage shows the unhappy test pilot, Vladimir Komarov, and some very subdued technicians. It was almost as if everybody knew that a bad day was in store. They had good reason to be anxious. Sergei Pavlovich Korolev,

Eisele (top left) and two shots of Cunningham making notes during the Apollo 7 mission. Notice the Hasselblad film magazine floating in the weightless interior of the spacecraft.

Apollo 7 commander Wally Schirra looks out the rendezvous window on the ninth day of an orbital mission, testing major spacecraft revisions in the wake of the Apollo 1 fire.

Walter Cunningham, Apollo 7's LM pilot. The spacecraft made a practice docking approach using a small disc-shaped target inside the depleted S-IVB stage's LM adapter.

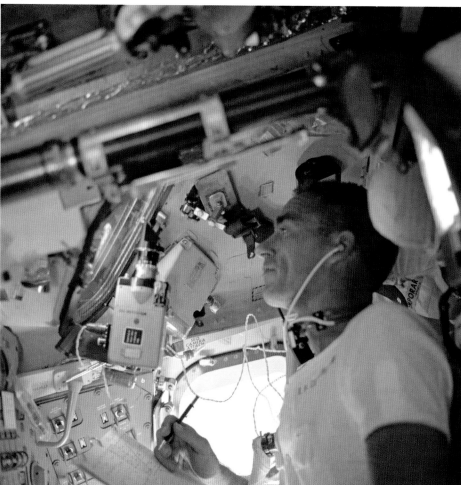

the man who made the Soviet space program such an epic success in its early years, had died. Weaker, less capable managers had taken his place. Yuri Gagarin and other cosmonauts had tried to send word to the Kremlin that the Soyuz was riddled with design flaws and should not be launched until it had been improved.

No one in authority listened, and just as his friends had anticipated, Komarov hit trouble as soon as he reached orbit. Launch of a second Soyuz that was supposed to rendezvous with Komarov was canceled while ground controllers worked to fix numerous power and control problems. During Komarov's reentry, the parachutes did not deploy properly. His capsule slammed into the ground with all the force of a 3-ton meteorite, and he was killed instantly. NASA and its astronauts sent messages of sympathy, just as the Soviets had done for the Apollo 1 crew. Skeptical lawmakers on Capitol Hill could no longer suggest that NASA had been unusually careless in its preparations. Space exploration was hazardous for both sides, and especially so in the febrile atmosphere of a race.

This need for context has not changed in the half century since the Apollo era. Rocket flight continues to be an unusually difficult and dangerous enterprise, and the vacuum of space will always present its own special difficulties. To date, just under six hundred people have flown into space: some once, many repeatedly. Of those six hundred, three (the crew of Apollo 1) were killed before their craft ever left the Earth, another (Vladimir Komarov) crashed to the ground, three cosmonauts (Russia's first space station crew) were suffocated in space, seven were blown apart soon after launch aboard NASA's *Challenger* shuttle, and seven more died in a high-altitude disintegration of *Columbia*. This is a fatality rate of one space traveler in twenty-eight. A risky profession indeed. Yet somehow the illusion persists that space travel should be a matter of routine, and that accidents ought to be unusual, and blameworthy. If we want to continue with it, we need to acknowledge the dangers, because we will never be able to eradicate them.

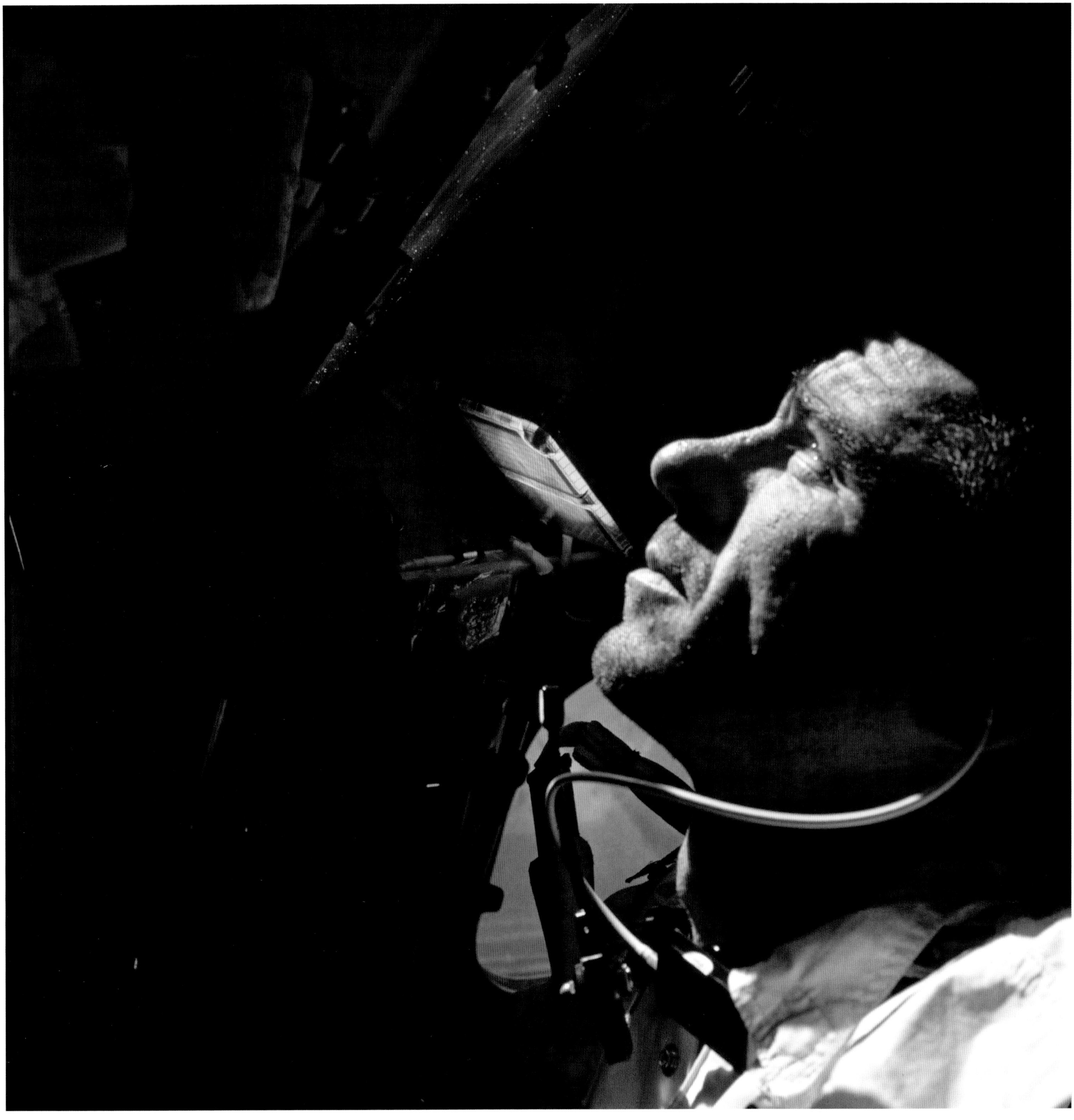

> IF ANY MAN WANTS TO ASK FOR WHOM THE APOLLO BELL TOLLS, I CAN TELL HIM. IT TOLLS FOR HIM AND FOR ME, AS WELL AS FOR GRISSOM, WHITE, AND CHAFFEE. IT TOLLS FOR EVERY ASTRONAUT OR TEST PILOT WHO WILL LOSE HIS LIFE ON SOME LONELY HILL ON THE MOON OR MARS. IT TOLLS FOR GOVERNMENT AND INDUSTRIAL EXECUTIVES AND LEGISLATORS ALIKE. IT TOLLS FOR AN OPEN PROGRAM CONTINUOUSLY EVALUATED BY OPINION-MAKERS WITH LITTLE TIME FOR SOBER THOUGHT, OPERATING IN THE BRILLIANT COLOR AND BRUTAL GLARE OF A WORLDWIDE MASS MEDIA THAT MOVES WITH THE SPEED OF A TV CAMERA FROM EUPHORIA TO EXAGGERATED DETAIL. WE HAVE A GRAVE RESPONSIBILITY TO WORK TOGETHER TO PURGE WHAT IS BAD IN THE SYSTEM WE TOGETHER HAVE CREATED AND SUPPORTED. WE HAVE PERHAPS AN EVEN GRAVER RESPONSIBILITY TO PRESERVE WHAT IS GOOD, AND WHAT REPRESENTS STILL, AT THIS HOUR, A HIGH POINT IN ALL MANKIND'S VISION.

NASA administrator James Webb, April 1968

The men of Apollo 1:
Virgil Grissom, Edward White,
and Roger Chaffee.

> **"THE APOLLO 1 FIRE WAS A CONSEQUENCE OF NOT KNOWING ENOUGH IN THOSE DAYS ABOUT SPACE FLIGHT. WE WERE ALL NOVICES . . . BUT OVERALL, THE RECORD HAS BEEN VERY GOOD."**
>
> **Apollo 9 commander James McDivitt, June 1999**

From inside their Command Module (CM) the crew of Apollo 9 photograph the Lunar Module (LM) prior to its extraction from the S-IVB upper stage.

Rusty Schweickart, wearing a Portable Life Support System (PLSS) backpack, explores the exterior of Apollo 9's LM.

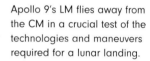

Apollo 9's LM flies away from the CM in a crucial test of the technologies and maneuvers required for a lunar landing.

Dave Scott emerges from the CM's hatch for a spacewalk, testing some of the spacesuit systems that will be used for lunar operations.

Having successfully jettisoned its Descent Stage, the LM makes a smooth rendezvous with the CM at the end of its free flight.

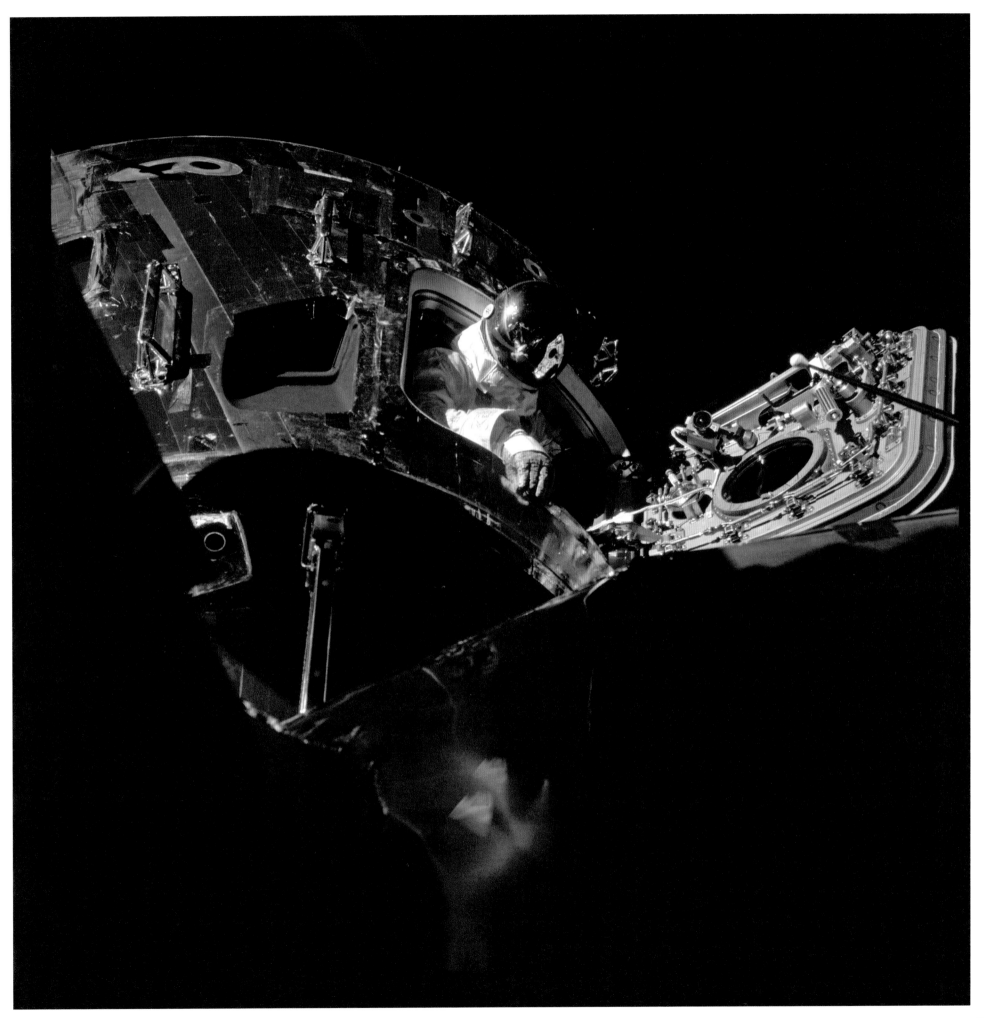

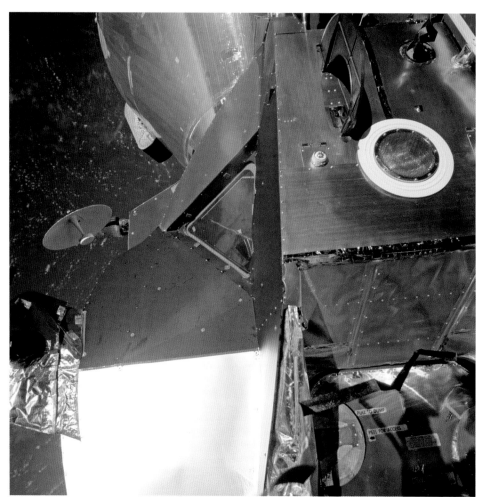

FROM THE EARTH TO THE MOON

McDivitt (top let) and Scott (bottom left) in good spirits during the Apollo 10 mission. A slightly out-of-focus shot of the CM's hatch (bottom right) shows the fast-opening mechanism, installed because of the dreadful delay in opening Apollo 1's hatch after the fire.

During the Apollo 1 investigation, a number of critics spoke disparagingly of NASA administrator James Webb's "Big Red Booster." He tried to warn Congress that the Soviets were building the N-1, a rocket almost as tall as a Saturn V. The CIA was monitoring the progress of this leviathan, but Webb kept the details vague, because no one was supposed to know about America's reconnaissance capabilities. He had a hard time convincing anyone, and it was only in the 1990s, after the collapse of the Soviet Union, that we in the West truly appreciated how enormous the N-1 really was.

The N-1's inaugural test flight in February 1969 was a disaster. A CIA satellite photographed the rocket standing on the pad, but next time the satellite flew overhead, the pad had vanished, and the landscape far around was scorched black. It soon became clear that the Soviets could not achieve their aim of landing a solo cosmonaut on the moon, either before Apollo or after it. However, a smaller and much more reliable rocket, known as the Proton, posed a more urgent challenge to NASA. On September 15, 1968, a Proton launched a mysterious unmanned spacecraft, Zond 5, and sent it around the moon. So far, so relatively routine. It was what happened next that really worried American observers. Zond's capsule, a pod large enough to carry a man, was returned to Earth, with its payload of biological samples (a tortoise and several plants) intact. Although the capsule's reentry was too hard and fast for human comfort, it looked as if the Soviets might try to steal a march on Apollo by sending cosmonauts around the moon and bringing them home. No doubt the flaws in the Soyuz vehicle could be fixed, just like Apollo's. If the Soviets had succeeded, it would have been a public relations catastrophe for NASA.

American observers saw outward signs of technical progress. What they did not see were the management failures and leadership rivalries slowing the Soviet's progress toward the moon. Consequently, NASA felt under pressure to get going. The Saturn V moon rocket had experienced major problems during unmanned test flights, so it was a bold decision to put astronauts on the next one, and send them to the moon, and into lunar orbit. With astronaut Frank Borman in command, accompanied by Bill Anders and Jim Lovell, Apollo 8 reached its target during the Christmas of 1968, at the end of a year that had been grueling not just for NASA but for America as a whole. The Vietnam War was escalating. Student riots, racial tensions, and civil unrest darkened the mood, as did the assassinations of civil rights leader Martin Luther King and presidential candidate Bobby Kennedy. Apollo 8's success delivered a much-needed sense of optimism about humanity's future.

Changing Our Perspective

Heading for the moon at 25,000 miles an hour, the crew of Apollo 8 sent back live TV pictures of the dwindling Earth: not just the curved horizon, but the entire ball of our planet, surrounded by the blackness of space. The monochrome TV camera aboard the ship revealed barely more than a white blob, but it was still sensational for people back home who had never seen anything like it. Jim Lovell thought out loud for his TV audience: "I keep imagining I'm a lonely traveler from some other planet, and I wonder what I'd think about the Earth from this distance. Would I think it was inhabited?"

Very soon after splashdown, and the worldwide celebration of the crew's safe homecoming, the Hasselblad film frames from the mission were processed and distributed. For the first time in history, we saw color images of the blue Earth suspended in the blackness of space, with the cold, lifeless, pockmarked lunar horizon serving as a vivid and urgent reminder that our little world is all we have. Looking back on his voyage a few years later, Apollo 8 crewman Bill Anders reflected, "I trained hard for many years to fly around the moon and take a close look at it, because I thought that's what the mission was. When we got home, I realized we had discovered something much more precious out there: the Earth. I believe the environmental movement was tremendously inspired by those missions. That alone is worth the relatively few billions of dollars that we spent on Apollo."

Tough Tests for Men and Machines

The launch rate in the Apollo era was astonishing by today's standards. On March 3, 1969, just two months after Borman and his crew had rounded the moon, Apollo 9 took off, carrying a launch module for the first time, nicknamed *Spider*. Commander James McDivitt and landing module pilot Russell "Rusty" Schweickart would take *Spider* for a test drive in Earth orbit, while command module pilot Dave Scott kept control of the mother ship, *Gumdrop*. The mission would also test what might happen if, for some reason, a command module and landing module could not redock at the end of lunar operations. Could future moon walkers bridge the gulf by making an emergency space walk?

Schweickart crawled out of *Spider*'s front hatch on March 6, while the vehicles were docked, but his ambitious schedule was simplified at the last moment because of the nasty attack of motion sickness he'd suffered a few hours earlier. Scott pushed his head through *Gumdrop*'s hatch, and both astronauts took spectacular photographs of each other. Schweickart recovered his equilibrium in time for *Spider*'s six-hour solo performance the following day, successfully carrying out almost all the maneuvers and rocket engine firings necessary for a lunar touchdown.

The space agency told a story of smooth progress toward the moon, and perhaps the public back home would have appreciated these routine test flights even more if it had been allowed to know just how tough they were. In his memoirs of 2004, Dave Scott observed, "NASA made those missions look too easy. They were really, really hard."

Apollo 10 took off on May 18, 1969, heading for the moon. Tom Stafford and John Young took their landing module, Snoopy, to within 10 miles of a touchdown. Watching through the windows of command module Charlie Brown, John Young caught a glimpse of Snoopy glittering in the sunlight 80 miles below him. When the time came for Snoopy's ascent stage to blast clear of the descent stage, and make the climb back to Charlie Brown's orbital altitude, the guidance computer experienced momentary confusion. The down-facing radar was processing data about the onrushing lunar surface, while the up-facing radar was hunting for Charlie Brown, and somehow, the two computer routines were trying to run simultaneously. Snoopy became schizophrenic, bucking and gyrating, literally not sure if it was trying to get away from the moon or drop down toward it.

"Son of a bitch!" Cernan swore. "We're in trouble," Stafford confirmed. They regained control, and the computer was successfully stabilized. After the mission, a pastor wrote to NASA, more concerned about Cernan's use of bad language than by the danger he'd been in. Cernan pleaded that an astronaut fighting a wayward machine could be forgiven for not always minding his language. "I saw the lunar horizon go by about seven or eight times in ten seconds, and that's a hair-raising experience," he said.

Assembled from adapted Redstone fuel tanks and other components, the Saturn 1-B rocket was used for lifting command modules into orbit.

A view south shows a large crater, Goclenius, in the foreground. The three craters at top right are Magelhaens, Magelhaens A, and Colombo A.

Last flown half a century ago, the Saturn V rocket still ranks as the largest and most powerful launch vehicle ever made. This photo shows an uncrewed test version on the pad during November 1967.

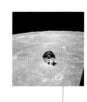
Apollo 10's CM seen from the LM as the two spacecraft maneuvered independently.

Earth rises over the lunar horizon, in these photos taken from inside Apollo 8's CM. This historic mission closely matched an idea first described by French author Jules Verne in his 1865 novel, "From the Earth to the Moon."

The LM's docking window, seen after the vehicle had descended to within 50,000 feet of the lunar surface, and climbed back up to rejoin Apollo 10's CM.

Apollo 8's view of a deep, darkly shadowed crater inside a larger crater 20 miles across, on the far side of the moon.

Apollo 10's oblique telephoto view of the lunar nearside shows Hyginus Rille, a dramatic feature more than 130 miles long.

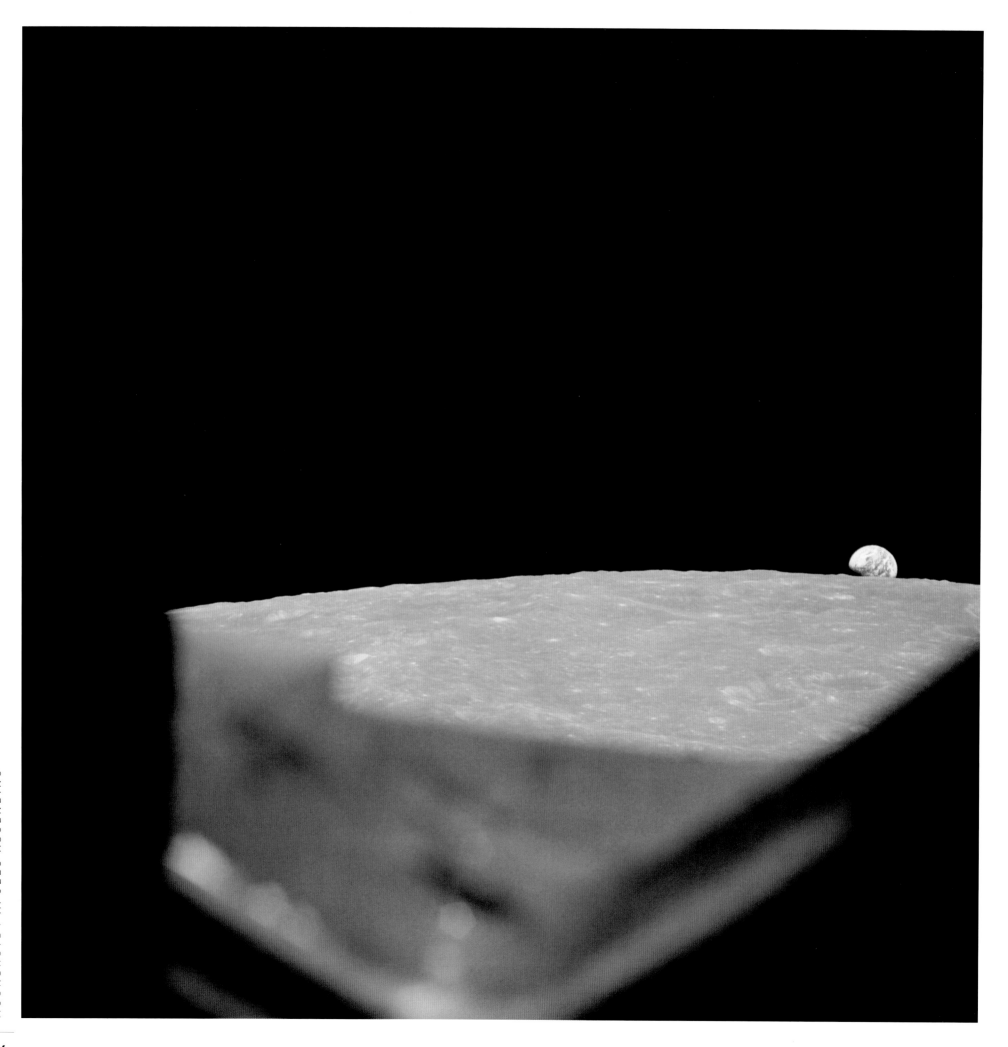

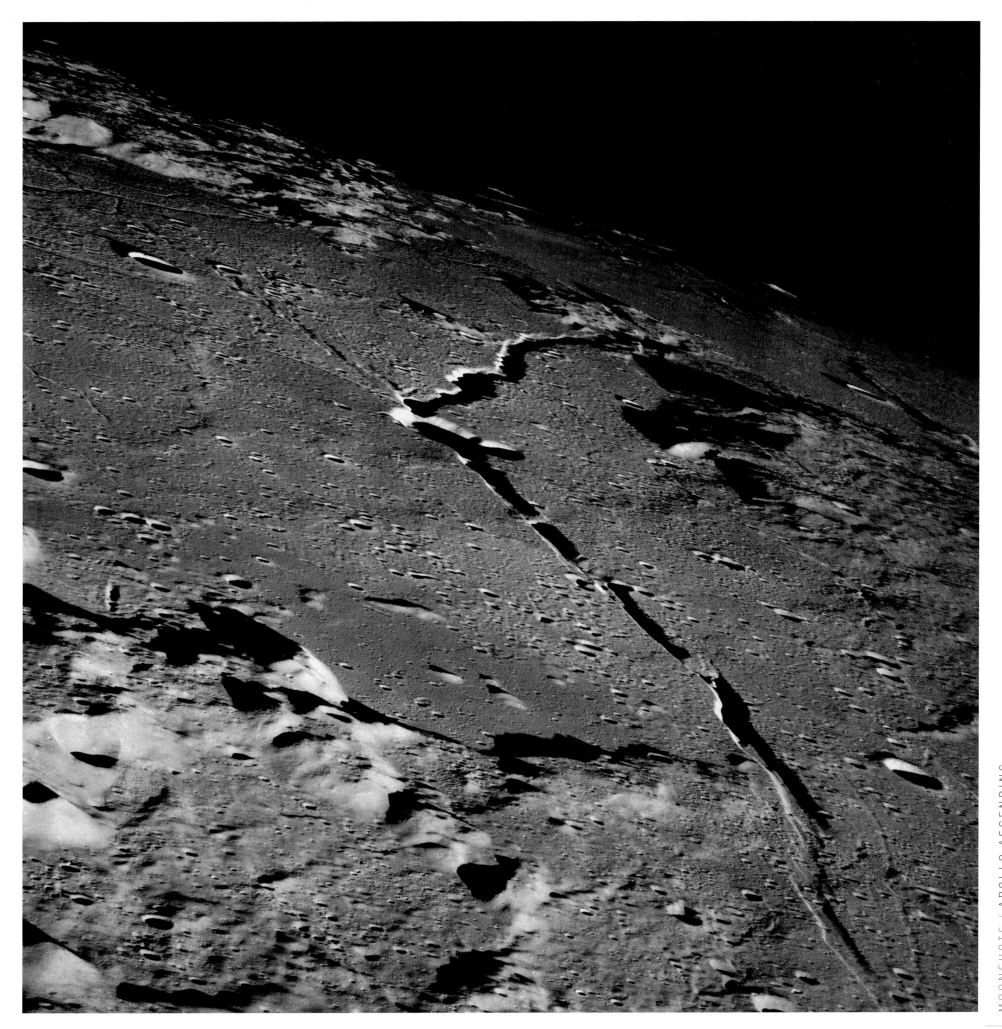

3 MAN ON THE MOON

In July 1969 human explorers stepped onto the surface of another world for the first time. We are still trying to absorb the meaning of this pivotal moment.

> "OUR AUTOPILOT WAS TAKING US INTO A VERY LARGE CRATER, ABOUT THE SIZE OF A BIG FOOTBALL STADIUM, WITH STEEP SLOPES COVERED WITH LARGE ROCKS ABOUT THE SIZE OF AUTOMOBILES. THAT WAS NOT THE KIND OF PLACE THAT I WANTED TO TRY TO MAKE THE FIRST LANDING."

Apollo 11 commander Neil Armstrong, 2005

On July 20, 1969, Neil Armstrong and Buzz Aldrin undocked Apollo 11's landing module, *Eagle*, and began their historic voyage down to the lunar surface. From this point, they could fly back up to rejoin Mike Collins aboard command module *Columbia* at almost any time, except during a brief period known by NASA insiders as the Dead Man's Zone. Within the final three minutes of the landing approach was a ten-second phase when *Eagle* would be descending so fast, no matter how hard it tried to "abort" by firing the ascent stage and climbing back up to lunar orbit, all its fuel would be exhausted working against the downward momentum. If anything went wrong during that ten seconds, Armstrong and Aldrin would be doomed to crash on the moon; and with the Dead Man's Zone less than half a minute away, something did appear to go wrong. A computer alarm lit up in the cabin. "Program alarm," Armstrong said. "1202," Aldrin confirmed.

At Mission Control, twenty-six-year-old Steve Bales was monitoring *Eagle*'s computer systems. Senior flight controller Gene Kranz demanded to know, "What's a 1202?" even as yet another code flashed on *Eagle*'s control panel. "Stand by," Bales replied. But Armstrong needed an immediate answer. "Give us a reading on the 1202 alarm," he said. In astronaut-speak, he was asking if he should abort the landing.

If Bales was going to recommend an abort, he had to do so immediately. After swiftly polling his back room advisors, he made the bravest decision of his life and spoke into his headset for Kranz and all the other controllers to hear. On the voice tapes that survive to this day, his voice is shaky. The fate of Apollo 11, and the lives of the two men aboard *Eagle*, was in his hands. "We're 'Go' on that alarm," he said. Kranz was startled, but—in keeping with the regime of absolute trust that applied within Mission Control—he said nothing to contradict his young colleague. The computer carried on working, and *Eagle* plunged safely through the Dead Man's Zone, much to Bales's relief.

Just as Bales was recovering from that scare, CapCom Charlie Duke radioed a terse warning up to the *Eagle*, hovering now on a plume of engine exhaust just above the surface. "Sixty seconds." In technical shorthand, he was saying to Armstrong and Aldrin, "You only have sixty seconds of fuel left in your tank. You better land that thing."

Kranz and his team then saw something even more worrying on their telemetry screens. Just when the landing module was supposed to be settling down into the lunar soil, it shot forward sickeningly fast. Had Neil and Buzz lost control of their ship?

Another agonizing ten seconds crawled by. At last, Aldrin radioed, "Contact light. Mode control to Auto. Engine Arm off." Those were the first words ever spoken by a human on another world, but history prefers to recall Armstrong's words a few moments later. "Houston, Tranquility Base here. The *Eagle* has landed."

Duke's response was almost lost among the whoops of delight. "Roger, Tranquility, we copy you on the ground. You got a bunch of guys about to turn blue. We're breathing again. Thanks a lot."

Armstrong was apologetic when he radioed to explain what had happened. "That may have seemed like a very long final phase, but the auto targeting was taking us into a crater with a large number of big boulders and rocks." With his fuel running critically low, he had needed precious extra seconds to nudge the ship forward until he could find a safe place to land. The media eventually realized the significance of the 1202 alarms, and there were many accounts of Armstrong switching off the computer and seizing manual control to steer the lunar module away from danger. In fact there was never any conflict with the computer, because the astronauts were always supposed to be able to choose the exact patch of ground they wanted to touch down on. That's why the lunar module had windows, after all. After the mission, Armstrong defended his "God-given right to be wishy-washy about where I was going to land."

It would have been impossible to land without the Apollo guidance computer (AGC). *Eagle* was balanced on a plume of engine thrust from a single nozzle, like an upright pencil poised precariously on a fingertip. A pistol-grip translation controller enabled Armstrong to steer the descent engine's nozzle, which was pivoted on gimbals nudged by electromechanical actuators. But those were under the AGC's control. Armstrong could not have maintained the balance of an essentially unstable vehicle without having his steering commands refined every one-tenth of a second intervals by the AGC.

As for the alarms: prior to the mission, a last-minute decision had been made to keep one of *Eagle*'s radars locked onto *Columbia* throughout the landing phase, just in case Armstrong and Aldrin needed to beat a hasty retreat from their landing attempt. Meanwhile, the forward dish would maintain a constant watch on the remaining distance to the lunar surface. Prior simulations had not revealed that this would create one additional task too many for the AGC to handle. On July 20, as *Eagle* sped toward the moon, the AGC decided that checking on *Columbia*'s range should be sent to the back of the priority list. It flashed "1202" and "1201" codes to warn that it was running at close to capacity, but not for one moment did it malfunction. No wonder the AGC's designers felt that their achievements in computing were too little understood by the media, and in their retirement and old age, how weary they must have become of hearing that the astronauts landed "despite" a wayward computer, and that you could navigate to the moon today with the microchip inside a child's digital watch.

The Mystery of the Moon

In the summer of 1999, the respected polling company Gallup discovered that only one in two Americans knew that Neil Armstrong was the first man to walk on the moon. Interestingly, it was older people who were most likely to have forgotten this fact. Younger people scored better. More disturbingly, at least one in ten Americans believe the landings were faked. In the United Kingdom, half the adult population says the same. Apparently there is something about Apollo that has created a cultural disconnect. Is it simply that we live in a more cynical age? Has the internet spawned too many conspiracy theories? As we have discovered in recent years, popular distrust of government enterprises is widespread, but the Apollo narrative brings its own unique brand of problems, and they are worth discussing for a moment. The "fake landing" stories reflect unease about Apollo's place in technological history. In 1886, Karl Benz patented the first automobile. Forty years later, American industry was manufacturing four million of them a year. The Wright Brothers flew the first human-carrying powered aircraft in 1903. Forty years later, commercial air travel and its darker cousin, aerial combat, were routine, and some people even had their own private planes. In the 1950s, a few large companies and institutions were using computers to do their payroll or solve complex equations. Forty years later, we all had one to play games on. And so it goes. The lunar adventure nourished countless areas of the economy in subtle yet important ways, speeding up developments in software, metallurgy, precision welding, and other profitable but unglamorous skills, such as personnel management, risk analysis, and information flow. But these benefits have not been so easy for most of us to identify. The more obvious impact that we might have expected, judged according to previous historical trends in machinery, are absent. We do not all have rockets, and we cannot all travel to the moon.

Strangest of all, NASA and the U.S. government tell us that it would be almost as hard, and nearly as expensive, to build a lunar spacecraft now as it was half a century ago. Naturally, some people mutter that if we could accomplish Apollo using primitive technology, then we should be able to do it all again today, and at less cost, what with all the modern microchips and materials at our disposal, not to mention the vast backlog of experience accumulated in NASA's brain bank. The fact that we can't do this suggests that Apollo must have been some kind of a miracle the first time around. In a sense, it was. A series of political and cultural circumstances unique to the early 1960s set the scene for Apollo, propelling it into existence ahead of its time. As a society, even as a species, we were not quite ready to draw down the moon. And yet, we did.

And just as soon as we got there, the world moved on to new, even fresher distractions, precisely because Apollo refused to fit into any pattern of progress or development that anyone could process, let alone build upon. Literally, we had been to the moon and did not know what to do next, because we had never really known why we went there in the first place. From the outset, Apollo was a grand gesture, a work of hubris, an act of faith, rather than a specific plan for technological or social development. As the American writer Norman Mailer observed in 1969, "For the first time in history, a massive bureaucracy had committed itself to a surrealist adventure. The meaning of the proposed act was clear to everyone, yet nobody could explain its logic."

In the 1960s, NASA was at the vanguard of a new technology. Today it has become the custodian of a rather old-fashioned one. Rockets are no longer the prime symbols of modernity, and the ballistic missiles that triggered the Space Age in the first place have lost much of their dark prestige. A handful of fanatics with ten-dollar craft knives can change the world in a day. Once-powerful nation states look on in confusion, while their expensive nuclear megatons sit silent in their underground silos, all but impotent. Where would they be aimed? What good could they do anyone now? If the missile seems redundant, no longer suited to the needs of a fragmented supra-national world, so too does the moon rocket, whose thunderous ascent on a pillar of fire was designed to impress a very different audience than today's.

And yet, none of these circumstances can detract from the glory of Apollo. It reached the moon in fulfillment of a dream that had enticed humankind for countless generations. The incredible scale and ambition of NASA in those years was an expression of political idealism from an America rather different to the one we know today: a nation that hoped for a better future rather than simply trying to protect itself from the perils of an uncertain present. Some people would like to revive that spirit of optimism now, both in space and on the ground.

There is more to human striving than usefulness, value for money, or indeed, logic. Some would argue that Apollo's greatest asset was its pointlessness. It has become an easy cliché to compare the tall white spire of the Saturn V to a cathedral, but the thing about clichés is that they usually have some truth in them. The great medieval castles were impressive statements of power and prestige, yet they were brutishly practical, and not nearly so much of a challenge to build—and of course, not so joyously beautiful—as the unnecessarily tall, and pointlessly ornate cathedrals, which were designed to show off what could be done with stone, wood, and glass when those materials were stretched to their limits. For all the practical reasons of church and state that ensured their funding, the simple truth is that cathedrals rose up from the ground because communities wanted to express the utmost that they could achieve. Perhaps Apollo was a similar act of faith made by the United States of America.

While we wait, ever hopeful, for the human spirit to do what it does best—constantly to reinvent and revive itself, and move on, and strive for betterment—we might recall what the novelist and pioneering space champion Arthur C. Clarke wrote in 1972, as the lunar missions came to an end: "An age may come when Project Apollo is the only thing by which most people remember the United States, or even the world of their ancestors, the distant planet Earth."

It might seem strange, but as Apollo 11 lifts off at the start of one of the greatest journeys in history, all the NASA launch controllers (overleaf) have their backs to the job is to watch their readout screens, and not be diverted by the spectacular sight of the rocket itself.

PAYING OUR RESPECTS TO YESTERDAY'S TOMORROW

When we look at the distinctly second-hand, half-century-old museum specimens left over from the days of Apollo, we wonder how anyone could have dared to reach for the moon aboard such primitive machines. When we peer through the hatches, we see dimly lit interiors encrusted with ancient clockwork. The instrument panels are a maze of switches and dials, and the thousand-fold electrical energies that once powered them are no longer in evidence to lend them some life.

To the generation who witnessed it in action, Apollo was incredibly futuristic, and then some. Picture the Command Module brand new, completely covered in mirror-smooth silver thermal insulation so that it looked like a single piece of polished metal. (The foil burned off during re-entry, leaving the capsule a drab brown color. It's best to think of it in its launch day finery.) Now visualize the interior of the spacecraft surgically clean and brightly lit so that the astronauts can see what they are doing. There is not a single scuff mark, nor the tiniest flaw in the paintwork. The life support systems units are humming, and the control panel is a shimmer of lights and trembling dials. Apollo and its insect-like Lunar Module seemed almost alive. By the standards of the 1960s, they were the most complicated objects ever conceived by the human mind. How will today's clever devices appear to the children of tomorrow, when they encounter their dusty remnants in a museum?

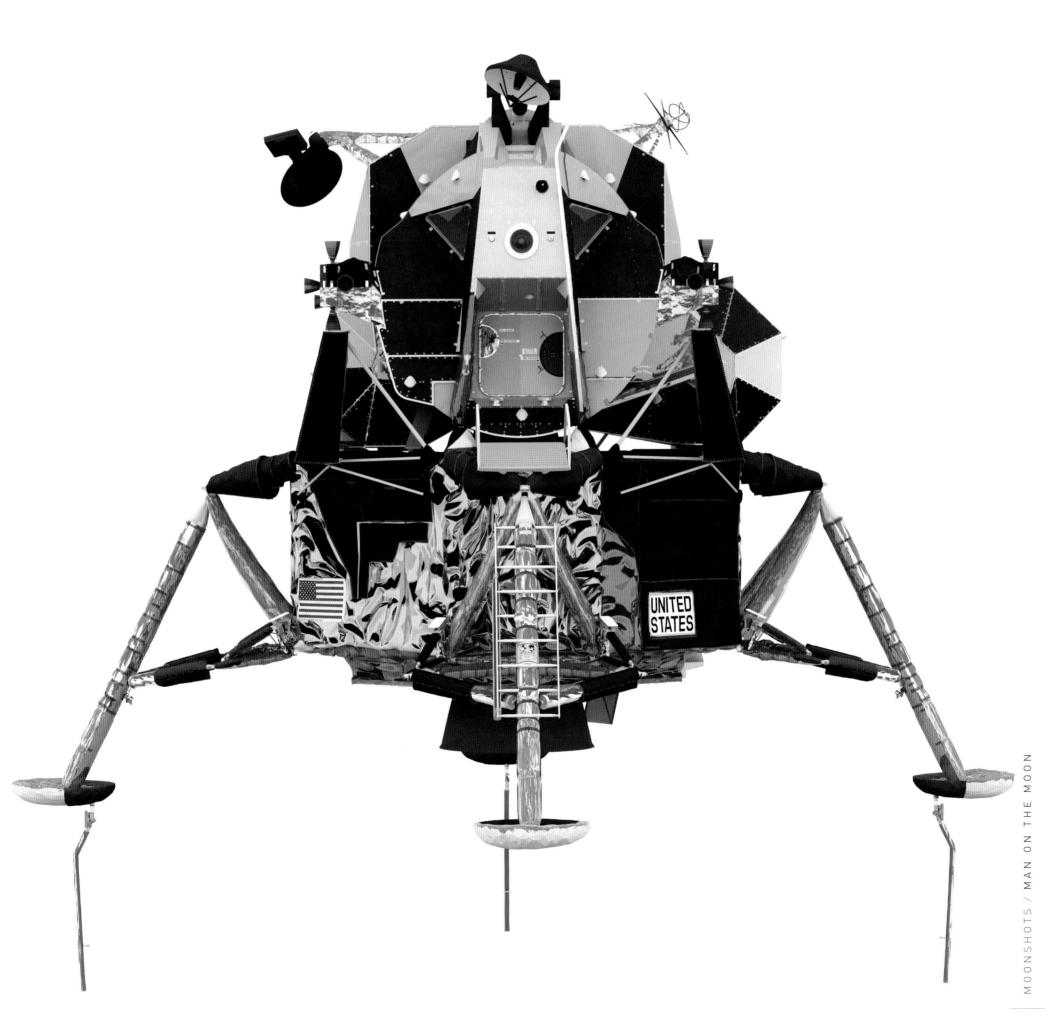

 The perfectly preserved interior of the Apollo 11 CM, housed in the Smithsonian National Air and Space Museum, Washington, D.C.

 A shot of Apollo 11's LM taken from the CM just as it approaches to dock with the LM and pull it clear of the S-IVB carrier stage.

 The interior of a LM originally built for an unmanned test flight, and subsequently used for ground tests.

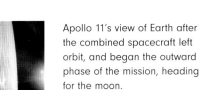 Earth dwindles as Apollo 11 speeds away and enters the realms of space dominated by the moon's gravity.

 An extreme telephoto shot of Apollo 11's Saturn V in flight, taken by a military-grade tracking camera on the ground.

 Apollo 11's view of Earth after the combined spacecraft left orbit, and began the outward phase of the mission, heading for the moon.

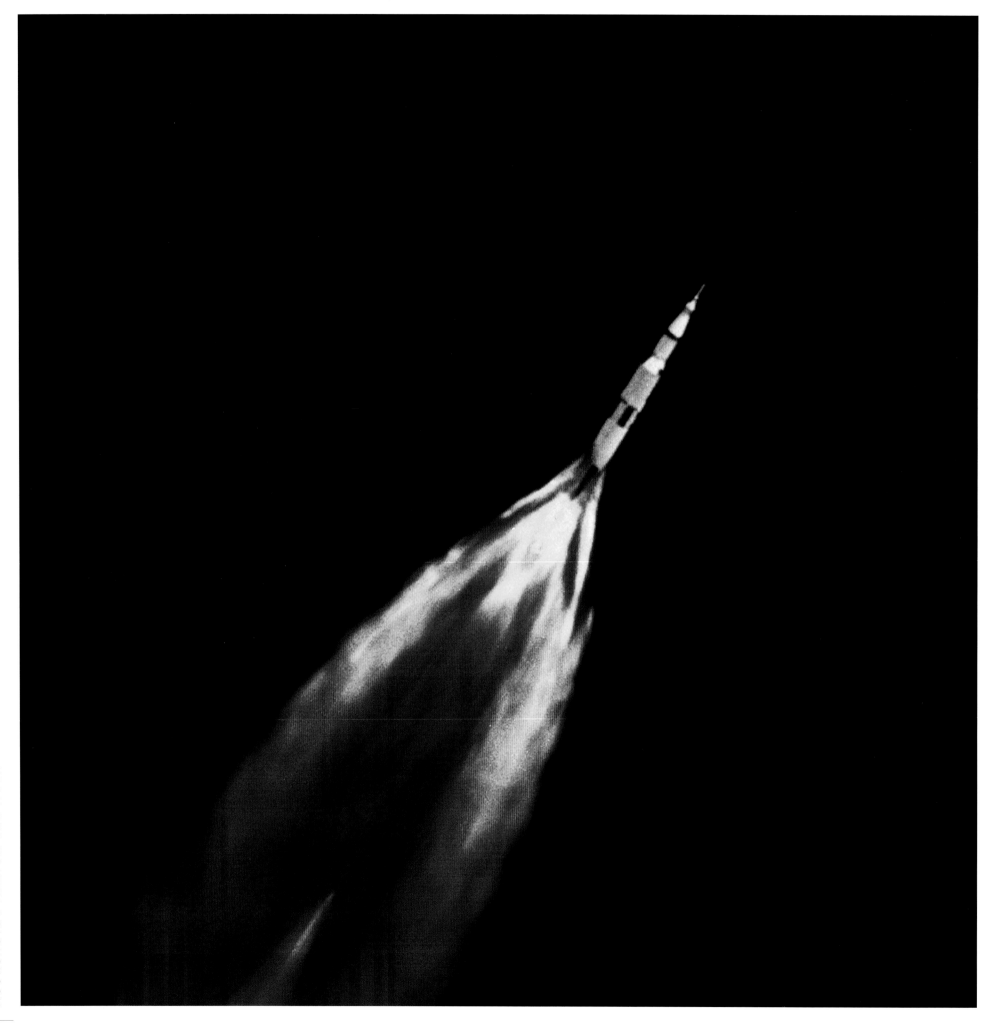

> "I REALLY BELIEVE THAT IF THE POLITICAL LEADERS OF THE WORLD COULD SEE THEIR PLANET FROM A DISTANCE OF 100,000 MILES THEIR OUTLOOK COULD BE FUNDAMENTALLY CHANGED. THAT ALL-IMPORTANT BORDER WOULD BE INVISIBLE, THAT NOISY ARGUMENT SILENCED. THE TINY GLOBE WOULD CONTINUE TO TURN, SERENELY IGNORING ITS SUBDIVISIONS, PRESENTING A UNIFIED FACE THAT WOULD CRY OUT FOR UNIFIED UNDERSTANDING, FOR HOMOGENEOUS TREATMENT. EARTH MUST BECOME AS IT APPEARS: BLUE AND WHITE, NOT CAPITALIST OR COMMUNIST; BLUE AND WHITE, NOT RICH OR POOR; BLUE AND WHITE, NOT ENVIOUS OR ENVIED."

Apollo 11 astronaut Michael Collins, July 2009

Five Hasselblad photos are combined to give an impression of LM pilot Buzz Aldrin temporarily occupying Armstrong's position while he checks out the LM's systems.

The LM sets off for its solo flight, and performs last-minute test maneuvers while it is still within visual range of the CM.

Neil Armstrong was the first man to step onto the lunar surface, but most of the famous Apollo 11 images are of Buzz Aldrin, photographed here by Armstrong as he descends the LM ladder and begins to work around the vicinity of the LM. The boot print is also Aldrin's.

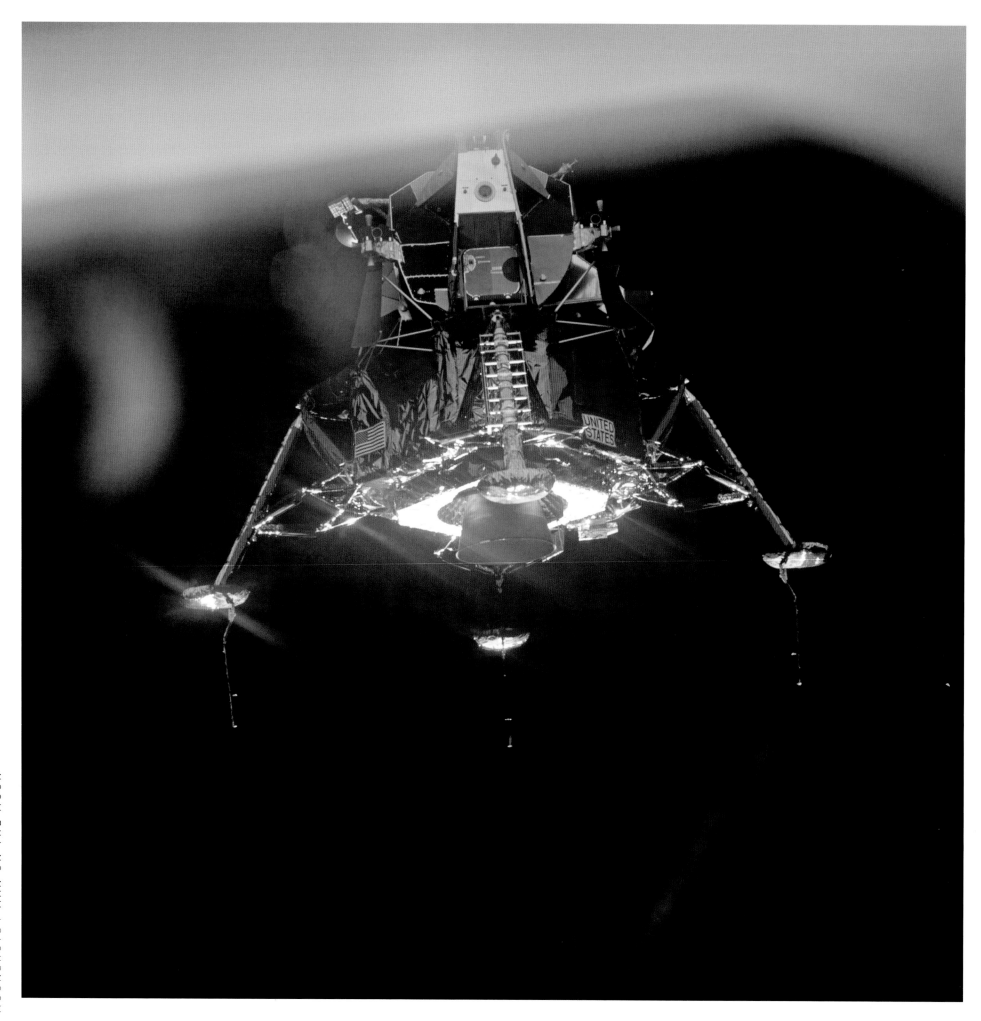

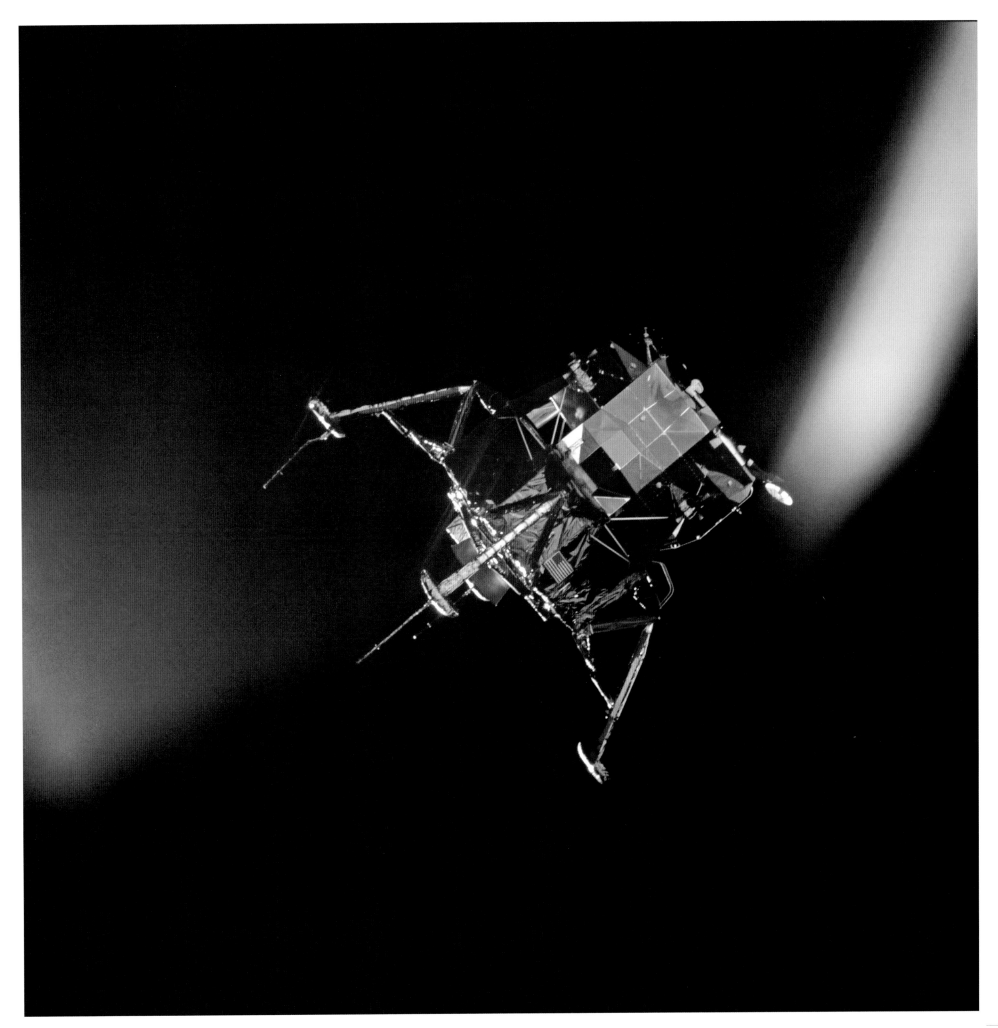

This composite image allows us a wider view of Aldrin's descent, as captured by Armstrong. All the moon walkers were trained to take overlapping shots that could be assembled later into panoramic views.

Aldrin's photo of Armstrong shows a happy Apollo 11 commander at the end of lunar surface operations. The next challenge would be to leave the surface safely, and rejoin Collins aboard the CM.

Apollo 11's orbital view of Daedalus, a 20-mile wide impact crater near the equator, on the far side of the moon. It was created by a massively powerful asteroid collision.

A view from the LM of the CM, almost lost among the rugged lunar backdrop. This view includes western areas of the Sea of Tranquility.

Fine views of the CM (top) and the LM's Ascent Stage in orbit above the moon, during a mission that probably will be remembered, one way or another, for as long as human beings continue to exist.

" HALF A CENTURY LATER, WE LOOK BACK ON WHAT HAS BEEN ACCOMPLISHED. OUR KNOWLEDGE OF THE UNIVERSE AROUND US HAS INCREASED A THOUSAND FOLD AND MORE. WE LEARNED THAT HOMO SAPIENS WAS NOT FOREVER IMPRISONED BY THE GRAVITATIONAL FIELD OF EARTH. PERFORMANCE, EFFICIENCY, RELIABILITY, AND SAFETY OF AIRCRAFT HAVE IMPROVED REMARKABLY. WE'VE SENT PROBES THROUGHOUT THE SOLAR SYSTEM AND BEYOND. WE'VE SEEN DEEPLY INTO OUR UNIVERSE AND LOOKED BACKWARD NEARLY TO THE BEGINNING OF TIME . . .

. . . THE GOAL IS FAR MORE THAN JUST GOING FASTER AND HIGHER AND FURTHER. OUR RESPONSIBILITY IS TO DEVELOP NEW OPTIONS FOR FUTURE GENERATIONS: OPTIONS IN EXPANDING HUMAN KNOWLEDGE. "

Armstrong to NASA on 50th anniversary of the agency, March 2008

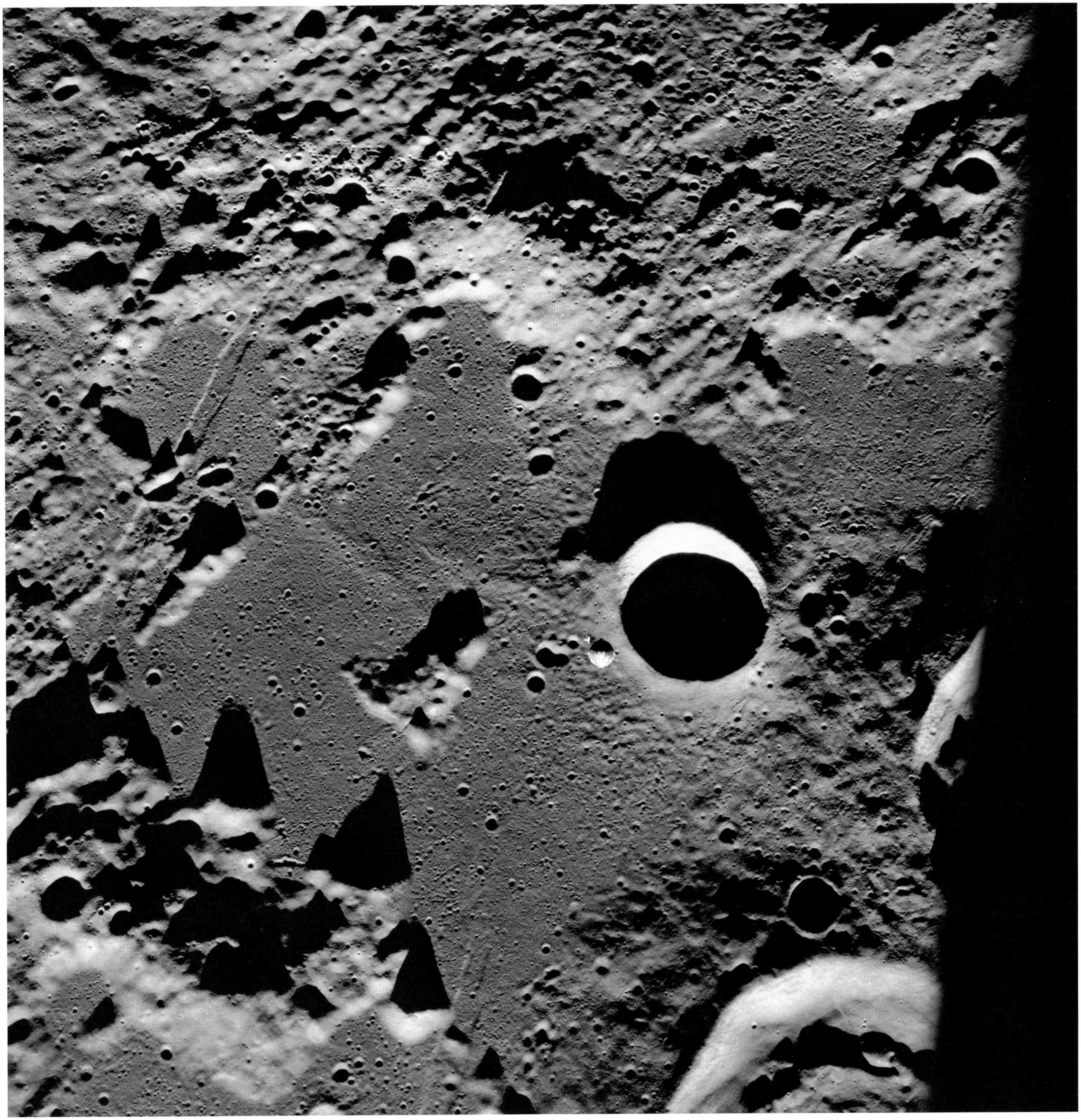

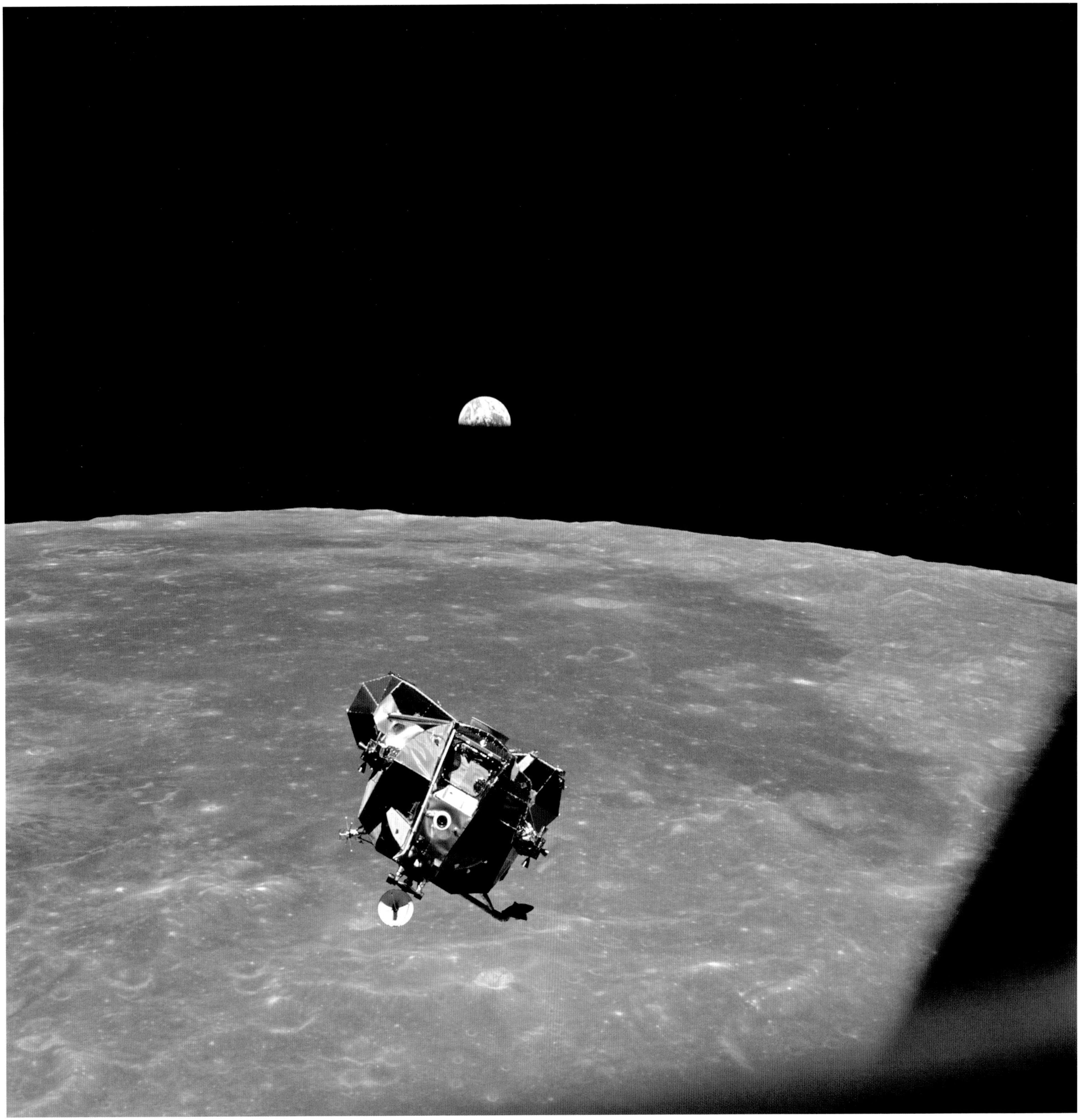

Apollo 12, at the top of its Saturn V launch vehicle, is rolled out of the Vehicle Assembly Building (VAB) for the first, slow phase of its Crawler-Transporter journey to the launch pad.

APOLLO 12
THE OCEAN
OF STORMS

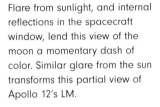

Flare from sunlight, and internal reflections in the spacecraft window, lend this view of the moon a momentary dash of color. Similar glare from the sun transforms this partial view of Apollo 12's LM.

Apollo 12 LM pilot Alan Bean as he prepares to step off the ladder to join mission commander Charles Conrad on the lunar surface.

Bean uses a core sampler to extract material from underneath the powdery topsoil of fine dust.

A fabulous portrait of Alan Bean. Notice the Hasselblad attached to his suit. Conrad's reflected image is visible in Bean's visor.

Apollo 12 touched down near NASA's unmanned Surveyor 3, which landed on the moon on April 19, 1967.

An exact duplicate of the Surveyor lander, originally used for engineering tests, and now on display.

The first men on the moon were a brilliant, very unassuming test pilot who hid from the limelight after his moment of greatness and a demon-haunted workaholic who took decades to recover from the emotional impact of his mission. There was never anything written in stone to say that Neil Armstrong and Buzz Aldrin were destined to play those roles. Random crew rotations and the availability of hardware were significant factors determining their fates.

The three astronauts aboard Apollo 12 command module *Yankee Clipper* were Pete Conrad in the commander's couch, command module pilot Dick Gordon, and lunar module pilot Al Bean. They'd known each other for many years, having been navy pilots together before joining NASA. Certainly this was one of NASA's happiest ships, although their mission got off to a worrying start half a minute after liftoff, when their vehicle was struck by lightning—twice—on the wet and cloudy late Floridian morning of November 14, 1969. Two massive bolts of electricity struck the tip of the Saturn V and traveled to the ground 3,000 feet below, all the way down the column of hot gases from the rocket's exhaust plume.

As Conrad recalled later, "I heard the master alarm ringing in my ears, and glanced over to the main instrument panel, and it was a sight to behold." Every warning light was on, turning the capsule's interior into something like a store window display at Christmastime. Mission Control was poised to call an abort, and the launch escape tower was primed to haul the capsule and its crew to safety, after which the Saturn V would be deliberately exploded by remote control to prevent any possibility of wayward pieces coming down to earth over a populated area.

A young electrical expert, John Aaron, suspected that the lightning had scrambled the signals transmitted by Apollo to Mission Control, but not necessarily harmed the actual systems inside the ship. From what he could see, the Saturn was still flying true. He suggested that the astronauts flip a particular switch on their control panel, somewhat as an internet help desk advises us to turn modems off and on again to flush glitches out of the circuitry. It worked, and readable data started streaming once more.

The *Yankee Clipper*'s guidance systems were only temporarily phased, and the Saturn rocket also contained its own system, which navigated Apollo 12 into precisely the correct orbit. After that, it was a simple matter for the astronauts to make sure that all the *Clipper*'s systems were back on line. We often hear that the on-board Apollo guidance computer (AGC) had less power than a modern digital watch. But it was certainly one of the toughest computers ever invented. It was a totally hard-wired system, with its software encoded as patterns of wiring, snaking in and out of the ring-shaped magnetic cores that could not be overwritten or erased. It shrugged off the lightning scare and booted back to life without fuss.

On November 19, Conrad and Bean in their landing module *Intrepid* landed with pinpoint accuracy on Oceanus Procellarum, near the rim of Surveyor crater, named in honor of another landing craft that even then was sitting on the moon patiently awaiting its first visitors.

As Conrad stepped onto the lunar surface on November 19 for the first of two moonwalks during this mission, he said, "Whoopee! Man, that may have been a small step for Neil, but it's a long one for me." Bean joined him and promptly destroyed the TV camera by accidentally pointing it at the sun. Not that it mattered. The public was less interested in this mission now that the excitement of Apollo 11 had faded.

During the second EVA (extravehicular activity) on November 20, Conrad and Bean reached one of their prime targets: the Surveyor III robotic lander, dusty but still intact after two years on the moon. They retrieved its camera for study back on Earth. NASA scientists later found that earthly bacteria had survived inside the casing. Apart from the scary liftoff, and a TV camera accidentally burned out by being pointed at the sun, Apollo 12 was a flawless mission. The press and the public started to believe that flying to the moon was just a matter of routine and, perhaps, a little dull. NASA's next lunar journey would reawaken everyone's interest with a vengeance.

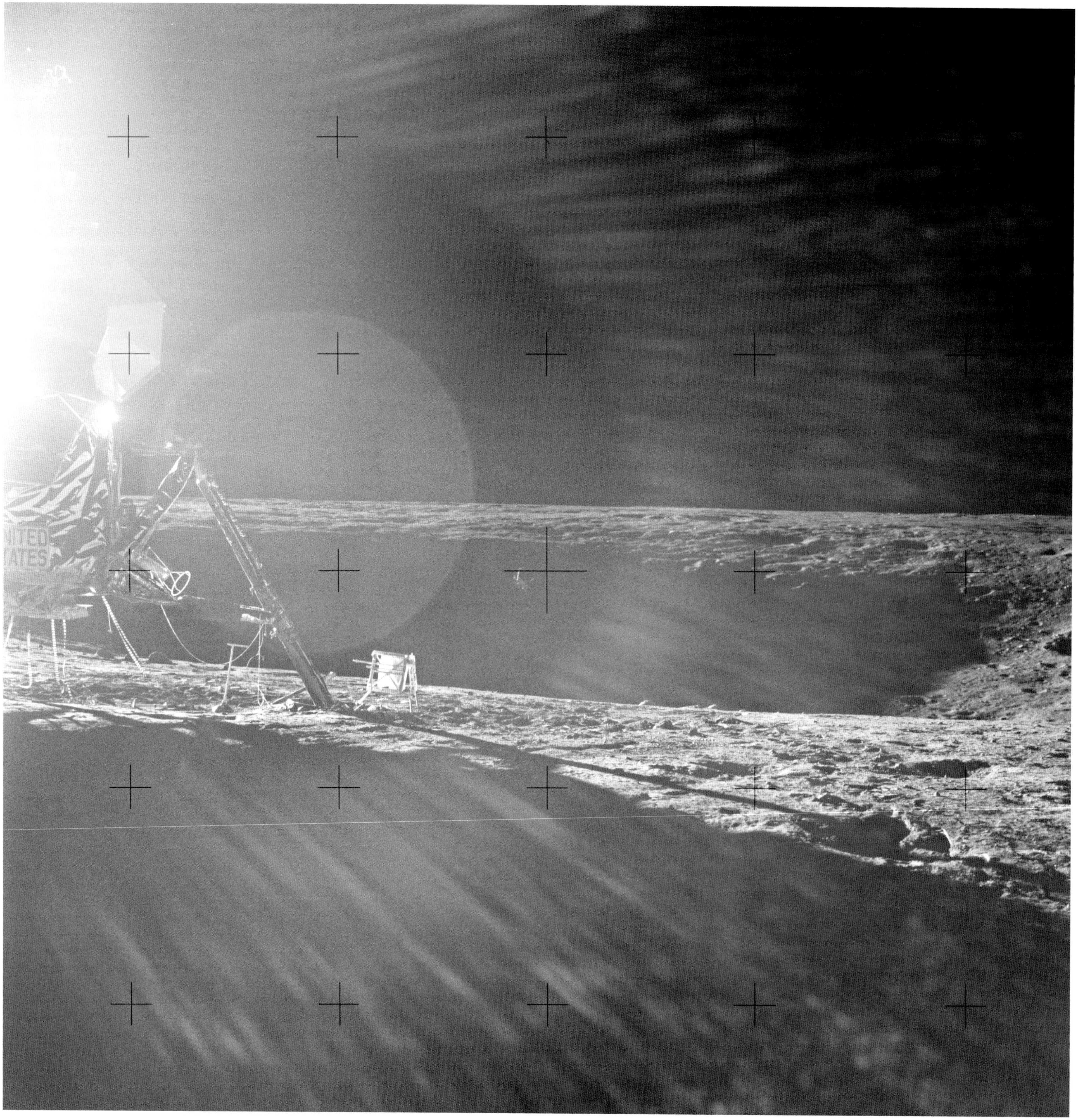
NITED
TATES

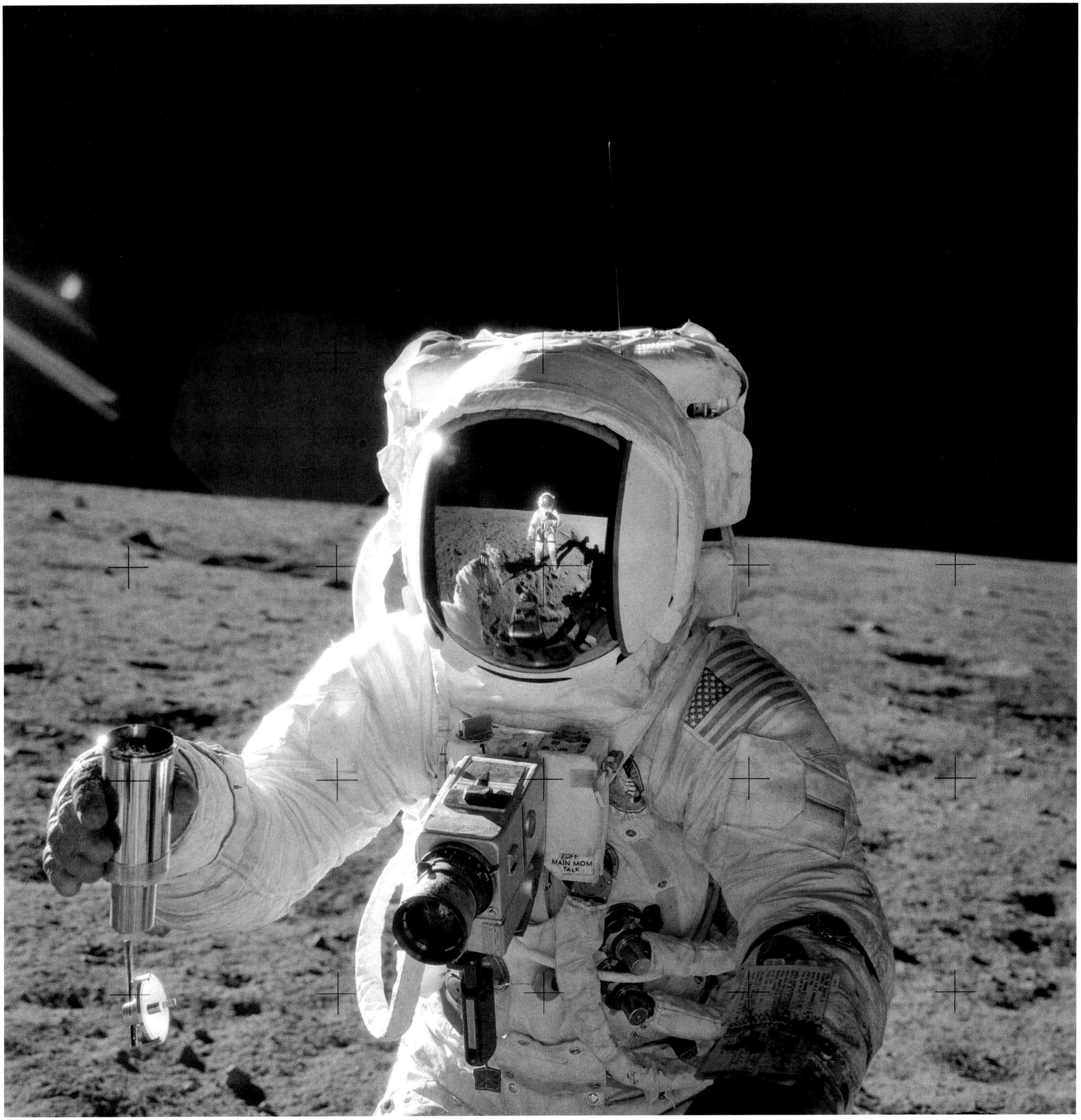
MAIN MOM
TALK

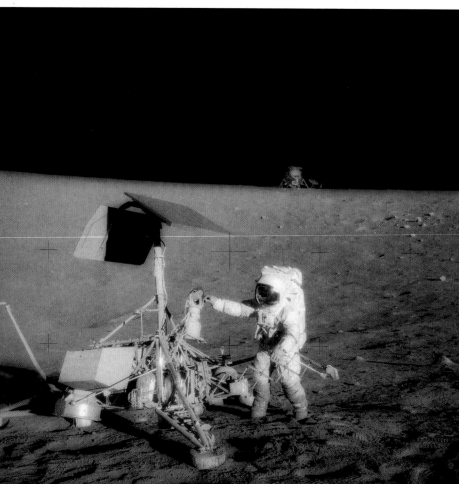

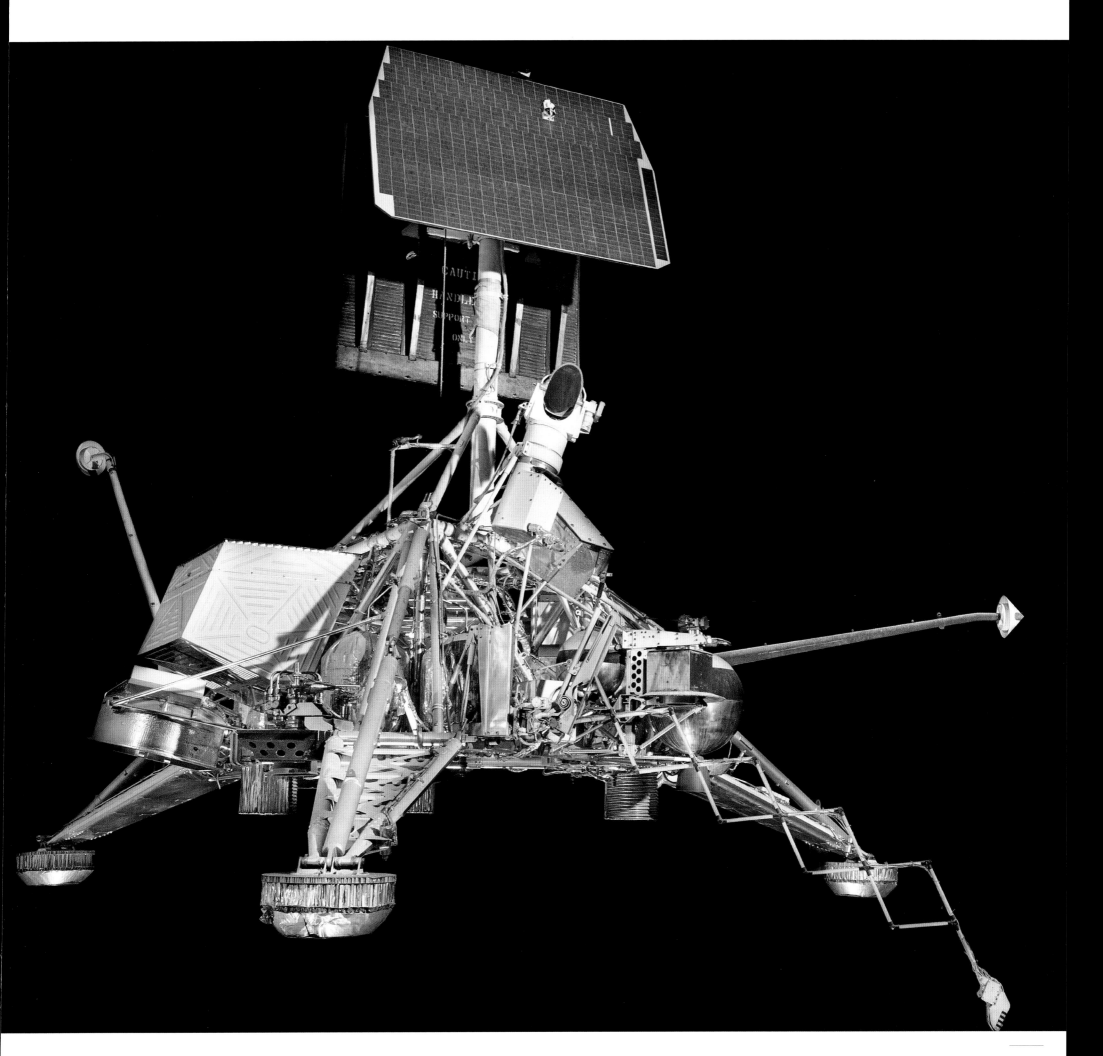

APOLLO 13
THE TRIALS OF
AQUARIUS

Apollo 13's gigantic Saturn V stands at Pad A, Launch complex 39, three weeks before lifting off, beginning one of the most dramatic space flights in NASA's history.

Apollo 13, launched at thirteen minutes past the hour, ran into trouble on the thirteenth day of April 1970, but the cause of the mission's famous problem was nothing to do with unlucky numbers. In the earliest days of spacecraft construction, the electrics had been modified to run with different voltages than originally planned. Among tens of thousands of components that needed to be very slightly adapted, a warming unit in an oxygen tank was overlooked. Five years later, it burned out while surrounded by liquid oxygen. Apollo 13 just happened to be the mission that suffered the consequences.

Commander Jim Lovell, lunar module pilot Fred Haise, and command module pilot Jack Swigert became household names because everybody back on Earth thought they were about to die. Everybody, that is, except mission controllers in Houston, who set to work on inventing one of the most brilliant rescues in space history.

At first, luck had seemed to be on Apollo 13's side. Five minutes after liftoff, one of the Saturn's second stage engines failed, but the other four took up the slack, burning for thirty seconds longer than usual. "There's nothing like an interesting launch," Lovell told mission control. As his ship entered a perfect orbit around the Earth, he added, "Houston, we're beginning to see a beautiful sunrise here." Lovell had rounded the moon as command module pilot aboard Apollo 8, which had not carried a lander. Now, as commander of this mission, he was eager to walk on the lunar surface. This dream was about to be snatched away. The first hint of trouble came fifty-six hours into the flight, when the spacecraft was far out from Earth and irretrievably on its way to the moon. Swigert radioed, "I believe we've had a problem here."

"This is Houston. Say again, please?"

Lovell confirmed that their command module, named *Odyssey*, was no longer a happy ship. "Houston, we've had a problem. We've had a main B bus undervolt." This was astronaut jargon for a major power failure. Haise added that "a pretty large bang" had accompanied the warning lights on their control panels. Lovell then peered through his side window. "Houston, we are venting something into space. It's a gas of some sort."

An explosion in the service module at the rear had destroyed the ship's oxygen supply, crippled most of its electrical systems, and knocked out the main engine. The astronauts switched off almost everything in the command module, saving what was left of its power reserves. They huddled in the lunar module *Aquarius*, trying to work out how to use its landing engine for an unexpected purpose. Instead of lowering two men gently down onto the moon, it would push three men, and the lifeless bulk of the *Odyssey*, around the moon and back toward the Earth.

If any problems had shown up earlier, Apollo 13 could have flown a Free Return Trajectory, in which gravity alone would have ensured the spacecraft's safe return. Prior to the explosion, the outward phase of the voyage had appeared to be running smoothly, and the ship's course had been altered to enable *Aquarius*'s touchdown at Fra Mauro, sacrificing the Free Return. Now the crew would depend on *Aquarius*'s engine to save them from being marooned in space.

A cloud of debris from the burst oxygen tank obscured the navigational telescope's view. Lovell and his crew had to use the sun as an emergency guide, making swift calculations with paper and pencil, before timing vital engine burns with a wristwatch. Five hours after the explosion, *Aquarius*'s landing engine was fired to nudge the combined spacecraft back onto a homeward trajectory. After rounding the moon, the engine was fired again, shaving ten hours from the return journey.

Aquarius's life support was supposed to keep two men alive for two days. Now, three men needed to survive on those reserves for twice as long. Shortage of oxygen was less of a concern than the buildup of exhaled carbon dioxide, which could only be dealt with by combining the lithium hydroxide air filters in *Odyssey* and *Aquarius*. But one was square-shaped, the other, round. Mission control had to invent a solution using spacesuit hoses, tape, and document covers. Step by step, they relayed instructions to the cold and exhausted astronauts.

Just before reentry, the crippled service module was discarded. "There's one whole side of that spacecraft missing," Lovell said, amazed at the extent of the damage. Incredibly, everything inside the command module still worked. *Odyssey*'s batteries had just enough juice to power its systems for reentry, and the crew made it home. Lovell may have been sorry not to reach the lunar surface, but his "failed" mission is regarded as one of NASA's finest achievements.

> "I'M COMING BACK DOWN THROUGH THE TUNNEL, AND SUDDENLY THERE'S A HISS-BANG! AND THE SPACECRAFT ROCKS BACK AND FORTH. I LOOKED AT FRED TO SEE IF HE KNEW WHAT CAUSED IT. HE HAD NO IDEA. I LOOKED AT JACK. HE HAD NO IDEA. AND THEN OF COURSE, THINGS STARTED TO HAPPEN."
>
> **Apollo 13 commander Jim Lovell, May 1999**

Jim Lovell (top right) was sad that he missed his chance to walk on the moon.

The crew of Apollo 13, exhausted but happy, on the deck of the recovery ship USS *Iwo Jima*.

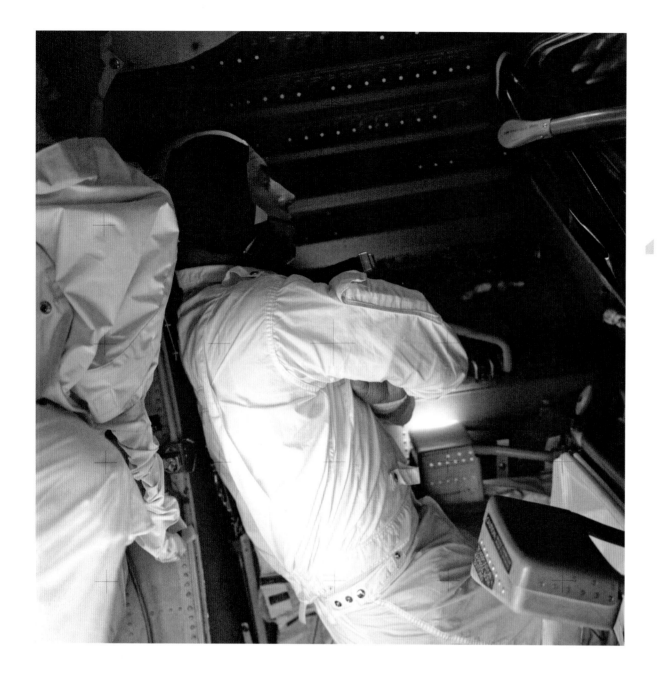

> " WE SAW OUR OXYGEN BEING DEPLETED. ONE TANK WAS COMPLETELY GONE. THE OTHER TANK HAD STARTED TO GO DOWN. THEN I LOOKED OUT THE WINDOW, AND WE SAW GAS ESCAPING FROM THE REAR END OF MY SPACECRAFT. IT DIDN'T TAKE MUCH INTELLIGENCE TO REALIZE THAT VERY SHORTLY WE WOULD BE OUT OF OXYGEN. THE THOUGHT CROSSED OUR MINDS THAT WE WERE IN DEEP TROUBLE. BUT WE NEVER DWELLED ON IT. "

Apollo 13 commander Jim Lovell, May 1999

Apollo 13 commander Lovell tries to get some rest in the LM, keeping warm as best he can in the frigid conditions.

Some of the emergency hose connections that became necessary when Apollo 13's crew moved out of the CM to use the LM as a "lifeboat." Jack Swigert is on the right.

A view of the moon from Apollo 13, with the CM visible at right, emphasizes how far from the safety of home the three astronauts were when disaster struck. Their recovery was a triumph of NASA ingenuity.

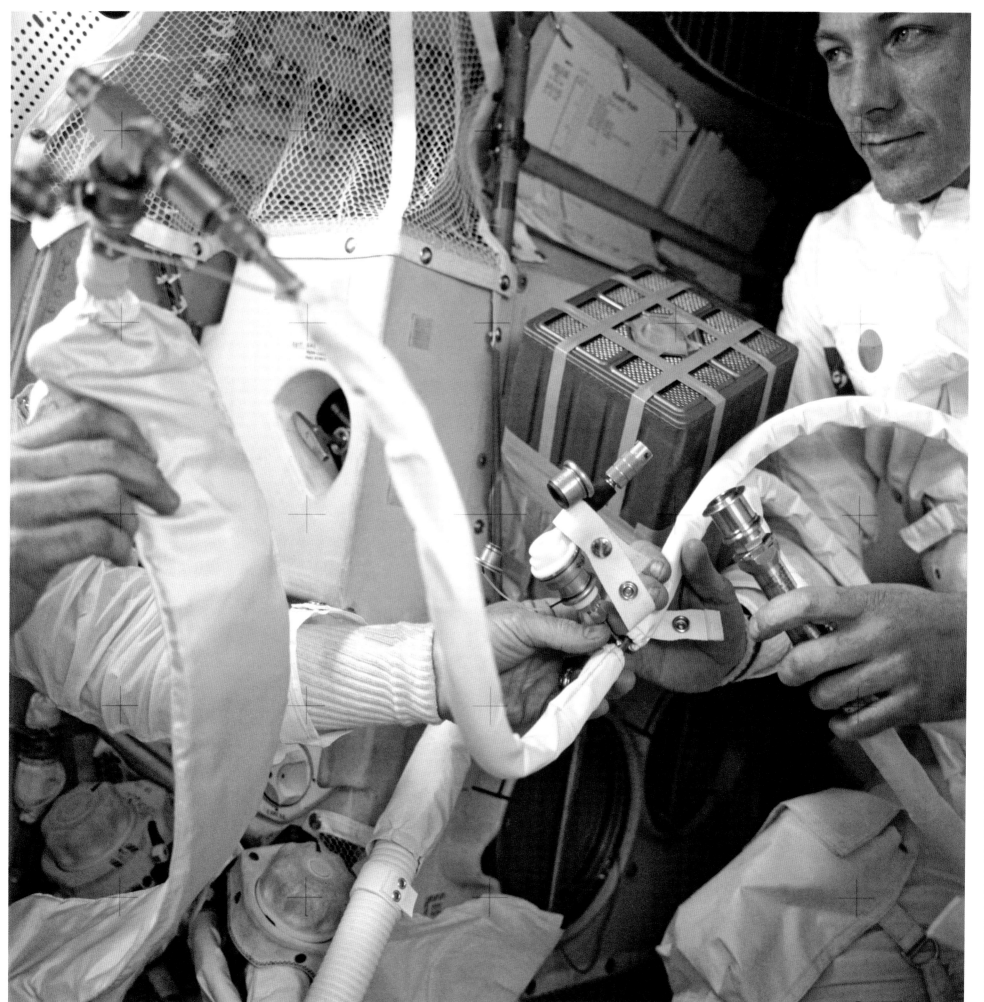

APOLLO 14
FRA MAURO

This unusual shot shows the mission commander's window of Apollo 14's LM, and its forward-facing rendezvous radar dish. A more complete view of this strangely shaped vehicle is centered on the docking collar.

LMs landed with the sun at their back, so that the astronauts' view of the approaching surface was not impeded by glare. Consequently, it was difficult to photograph an LM's front face after landing. As seen by Mitchell from inside the LM, Alan Shepherd raises his gloved hand against that same glare.

A composite view of the LM, lit perfectly by the sun. The right-hand landing leg has slipped slightly into the rim of a small crater, disturbing some topsoil.

Apollo 14 LM pilot Ed Mitchell poses by the U.S. flag. He was one of several Apollo astronauts who strongly sensed a spiritual significance in his journey.

Shepard handles a core sampling tube as he stands beside the Modularized Equipment Transporter (MEP), a wheeled handcart unique to Apollo 14. The sun highlights the trails of compacted lunar soil left by the MEP's wheels.

Apollo 14 was launched on January 30, 1971. On board were mission commander Alan Shepard, making his second space flight since his historic trip as America's first man in space back in 1961, and two rookies, lunar module pilot Ed Mitchell and command module pilot Stuart Roosa.

Problems with Shepard's inner ear had grounded him throughout most of the 1960s, but he had used that time constructively, serving as chief of NASA's astronaut office and putting himself through medical checks to regain his flight status at the grand old age (for an Apollo astronaut) of forty-seven. The last thing he wanted now was for any Apollo 13–style incidents to block his path to the moon.

Unfortunately, three hours into the mission, with the Saturn S-IVB upper stage and Apollo modules heading away from Earth and toward the moon, it looked as if Shepard might have his hopes dashed. Roosa couldn't get the command module *Kitty Hawk* to dock with landing module

Antares and pull it clear of the Saturn's S-IVB upper stage. Five times, he eased the command module's docking probe onto the roof of *Antares*, but it would not grip.

Mission control advised Roosa to push slightly harder, using a little more thruster force than normal. The sixth attempt worked. Later on, when the crew opened the hatches and looked at all the docking mechanisms, nothing seemed wrong. They had to assume that the next time the ships had to unite, there would be no similar problems . . .

Antares, with Shepard and Mitchell aboard, also had some unsettling moments during its descent to the lunar surface. The guidance computer indicated that an automatic abort might be triggered at any moment. In a tense rerun of similar alarms during Apollo 11's landing, mission controllers judged correctly that the warnings could be ignored. *Antares* touched down safely in the Fra Mauro highlands, the same area that was to have been explored by Apollo 13. Shepard and Mitchell conducted two EVAs totaling more than nine hours, using NASA's first pair of lunar wheels on a modular equipment transporter (MET), a hand cart loaded with sampling tools and cameras.

The local terrain was littered with debris from the ancient asteroid impact that created the vast Imbrium Basin. Fra Mauro itself was a smaller byproduct of that ancient chaos. Heat and pressure from the smash had fused countless rock fragments into new aggregates. In some cases, these had been smashed and reformed yet again by other impacts. Apollo 14 reinforced our view of the moon as a small world with an immensely violent history.

A few minutes before clambering back up the lunar module's ladder for the last time, Shepard gave mission control a surprise. Wielding a handle for one of the rock sampling tools, he announced, "It just so happens to have a genuine six iron on the bottom of it. In my left hand, I have a little white pellet that's familiar to millions of Americans. I'll drop it down." He then indulged in his favorite sport, becoming the first person to play golf on the moon, albeit one-handed, because "the suit is so stiff, I can't do this with two hands, but I'm going to try a little sand trap shot here."

His first shot simply gouged out a puff of lunar dust. Mitchell said, "You got more dirt than ball there, Al." The second attempt wasn't much good either. Back at mission control, Apollo 13 veteran Fred Haise in the CapCom seat lost no time in having a gentle dig. "That looked like a slice to me, Al." The third shot was better. "Here we go. Straight as a die," Shepard said. "Miles and miles and miles." Actually the ball traveled no more than about 600 feet, but Shepard, at long last, was happy.

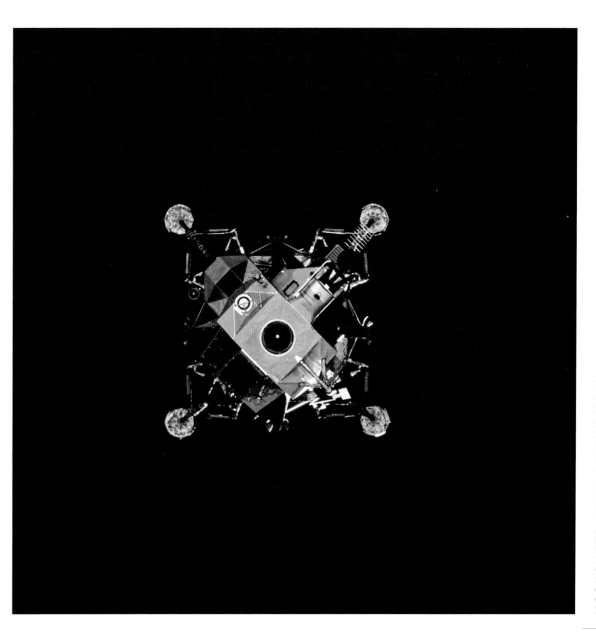

" EVEN THOUGH WE MIGHT HAVE TALKED IN TECHNOLOGICAL, FINANCIAL, AND POLITICAL TERMS, THE REAL PURPOSE WAS TO FIND OUR PLACE IN THE LARGER SCHEME OF THINGS. "

Apollo 14 astronaut Ed Mitchell, September 1996

Apollo 15 LM pilot James Irwin works by the LRV during the first period of lunar surface activity at the Hadley-Apennine site. The LM's shadow is in the foreground. Mount Hadley is in the background.

4 MAKING TRACKS

Apollo 15, 16, and 17 took advantage of a more powerful LM capable of delivering a wheeled Lunar Roving Vehicle to the surface of the moon.

Falcon, the landing module for Apollo 15, was a more powerful and heavily laden machine than its predecessors. It supported mission commander David Scott and landing module pilot James Irwin for three days on the lunar surface and carried nearly 4,000 pounds of scientific and exploratory equipment besides. Touching down with this additional load on July 30, 1971, was no walk in the park. Scott reported, "Okay, Houston, *Falcon* is on the Plain at Hadley." Irwin commented, "There's no denying that. It was a hard bump."

Among *Falcon*'s cargo, stored in the sides of the descent stage, was a new vehicle for space exploration, a four-wheeled battery-powered car known as the lunar roving vehicle (LRV). In the course of three drives, Scott and Irwin explored 16 miles of terrain, backdropped by the Apennine Mountains and the dramatic landscape surrounding Hadley Rille, a deep, lava-gouged groove that meanders for 60 miles at the foot of the Apennine range, along the eastern edge of Mare Imbrium (the "Sea of Rains").

During the second LRV excursion, while exploring Spur Crater on Mount Hadley Delta, Scott and Irwin retrieved what became known as the Genesis Rock. Early in the moon's 4.5 billion year history, just as its broiling surface of fluid magma was beginning to settle, some of the lighter materials floated to the surface, eventually cooling and solidifying into a hard crust. Volcanic instability consigned most of this crust to further transformations, but the Genesis Rock turned out to be a fragment of that original terrain, unaffected by subsequent lava flows. Geological training had taught the astronauts to expect that these highland regions probably would yield older rocks than the lava-swamped lowlands, but even so, they were delighted to find this one. Subsequent investigations on Earth revealed that the Genesis Rock's crystalline materials contain traces of hydroxyls, compounds of hydrogen and oxygen similar to water. This is mysterious, because our best theories about the moon tell us that it was created when two planetessimals, early bodies in the solar system, smashed together at colossal speed. The swarm of wreckage eventually coalesced under gravity to build a new world. The heat produced by this cosmic violence should have driven off lightweight, volatile gases such as hydrogen. Its presence in Genesis suggests that we do not know the full story of how our celestial neighbor was born.

Mapping the Moon

Orbiting overhead in command module *Endeavour*, Apollo 15's third man, Al Worden, also had some new equipment to run with. At first glance, the large rectangular space opened up on the service module behind him recalled the damage caused by Apollo 13's explosion, but this time the gap in the cladding was deliberate. A panel was jettisoned to reveal the scientific instrument module (SIM) bay, containing a suite of mapping cameras capable of resolving lunar surface details as small as 3 feet across, a laser altimeter for measuring the moon's contours with unprecedented accuracy, and a variety of instruments working in nonvisible realms of the spectrum. On August 5, while *Endeavour* was coasting back to Earth with all three crewmen safely aboard, Worden clambered outside the spacecraft to retrieve the film magazines from the SIM bay and bring them into the command module for return to Earth.

Irwin gives a military salute while standing beside the U.S. flag. Three miles behind him, Hadley Delta rises 13,000 feet above the plains.

Apollo 15 commander David Scott seated in the Lunar Roving Vehicle (LRV) during the mission's first lunar surface session.

" WHEN I LOOK AT THE MOON, I DO NOT SEE A HOSTILE, EMPTY WORLD. I SEE THE RADIANT BODY WHERE MAN HAS TAKEN HIS FIRST STEPS INTO A FRONTIER THAT WILL NEVER END. "

Apollo 15 commander David Scott, July 1971

A panoramic composite shows Apollo 15's LM backdropped by the Apennine mountains to the left and Hadley Delta Mountain on the right. The object next to the U.S. flag is solar wind particle collector.

A close-up view of a rock-strewn portion of a "relatively fresh" crater, looking southeast. The Apennine Front is in the left background, and Hadley Delta is visible to the right.

Apollo 15's Service Module carried a suite of instruments, including mapping cameras, one of which obtained this clear view of Hadley Rille.

Scott works alongside the LRV during the third and final period of Apollo 15 lunar surface activity. The LRV is parked a short distance from the rim of Hadley Rille.

Sunlight casts dramatic shadows, revealing the moon's violent, heavily cratered history, as Apollo 15 leaves that region of space and heads for home.

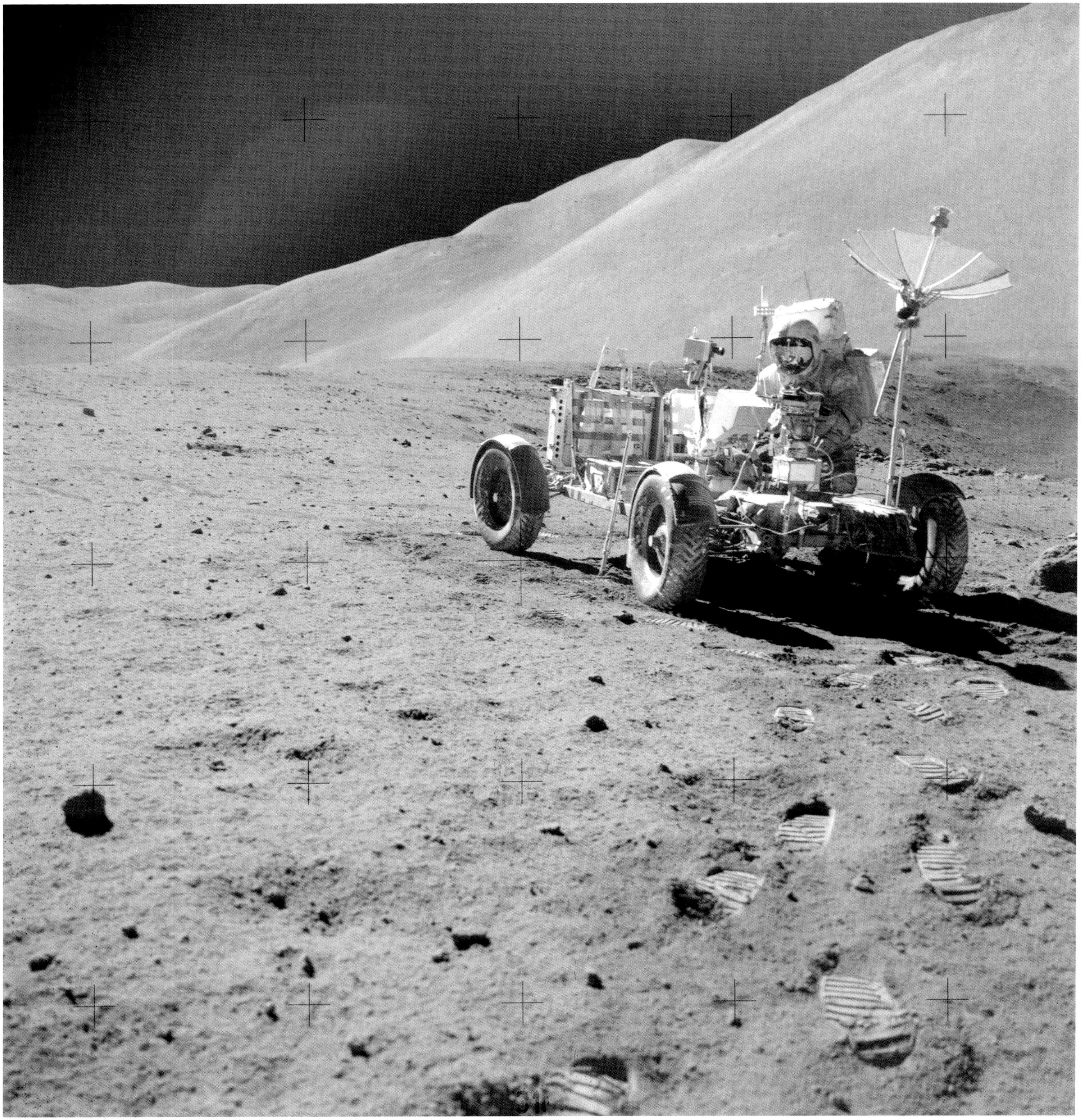

MOONSHOTS / MAKING TRACKS

An artist's impression of the spacewalk conducted by Apollo 15 CM pilot Al Worden to retrieve film cartridges from the lunar mapping cameras.

A view from the LM showing Apollo 15's CM. A panel has been jettisoned, exposing the lunar mapping cameras and other survey instruments.

A map, prepared in July 1971 by the United States Geological Survey, showing the journeys completed by Apollo 15's Lunar Roving Vehicle (LRV).

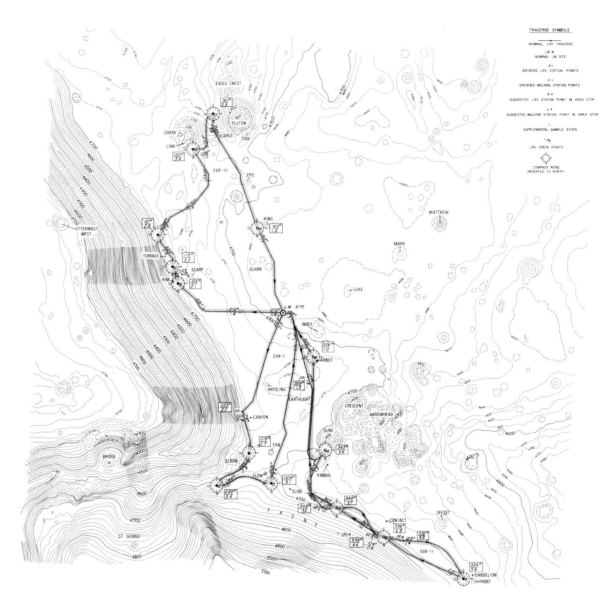

" BY THE SIDE OF THE SPACECRAFT, I DIDN'T DARE LOOK TOWARD THE SUN, KNOWING IT WOULD BE BLINDINGLY BRIGHT. AND ALL AROUND ME, THERE WAS NOTHING BUT BLACK. IT'S A SENSATION IMPOSSIBLE TO EXPERIENCE UNLESS YOU FLOAT TENS OF THOUSANDS OF MILES FROM THE NEAREST PLANET. "

Apollo 15 CM pilot Al Worden, 2011

APOLLO 16
DESCARTES
HIGHLANDS

Launched on April 16, 1972, Apollo 16 was the second mission to carry a lunar roving vehicle (LRV), stored in the descent stage of the landing module, *Orion*. John Young was mission commander, Ken Mattingly was pilot for command module *Casper*, and Charlie Duke took the right-hand position aboard *Orion*. Young and Duke's target was the Descartes lunar highland region.

Shortly after *Orion* undocked from *Casper*, the two spacecraft drifted in formation, undergoing final checks before *Orion* began its descent. Mattingly prepared to fire his main engine and push *Casper* into a suitable orbit for rescuing his colleagues in case anything went wrong with *Orion*. In fact, it was *Casper* that played up. As Mattingly checked backup control circuits for steering the service module's main engine bell, the entire spacecraft shuddered. "I be a sick bird," he said, clearly concerned that the landing attempt was endangered.

The two spacecraft continued to fly in close formation for another six hours until Mission Control decided it was safe for Mattingly to fire his engine so long as the primary steering controls were in order. They worked, and *Orion*'s solo voyage resumed. Young and Duke had been awake for thirteen hours by the time they landed and began their lunar visit with an emergency nap.

Apollo 16 commander John Young jumps and salutes the flag at the Descartes landing site. The LM is on the left, and Stone Mountain dominates the background.

Composite views of John Young working alongside the LRV, and ajusting the communications antenna so that it points more accurately toward the Earth.

Charlie Duke next to the LRV near Stone Mountain. The object in the foreground is a gnomon, a reference for determining true colors on the moon.

Duke near the steep rim of the 100-foot-diameter Plum Crater, with the LRV parked behind him.

The first of three EVAs was devoted mainly to unpacking the LRV and deploying the Apollo lunar surface experiment package (ALSEP). Young had trouble with the heat flow experiment, two probes pushed into the topsoil to detect faint warmth escaping from the moon's interior. As he walked away, a cable snagged on his boot and was wrenched free. He was more successful with an explosive "thumper," a rod with a cylinder at its end, shaped like a tin of beans, loaded with twenty-one explosive charges, each the size of a bullet casing. He pushed the business end against marked

points on a long sensor cable trailing on the lunar ground and fired a charge onto each marker, creating miniature seismic shocks that could be measured.

A mortar box with four rocket grenades created more powerful shocks. Each rocket was primed to travel different distances, ranging from 300 to 3,000 feet, and explode on contact with the ground. They were triggered by remote control a month after Apollo 16's departure (although the fourth launch was canceled after the third one tipped the mortar on its side).

Duke's problem was orange juice. The medics urged the astronauts to top up their potassium levels by drinking regularly from a tube dispenser inside their helmets. A leak soaked Duke's suit. He even got the sticky drink in his hair. Young complained too. "I like an occasional orange, I really do. But I'll be damned if I'm gonna be buried in them." Mission Control reminded Orion's crew to watch their language.

Looking back at this mission, Young regretted tripping over the heat sensor cables, but as a pilot, he was glad he had landed Orion with a healthy 100 seconds' fuel margin.

Even so, when he and Duke had climbed out of the landing module for their first surface session, they were alarmed to see a frighteningly deep crater just a few feet from *Orion*'s rear footpad.

The landing site also tripped up NASA geologists, because the astronauts could not see much evidence of the volcanic materials everyone had been expecting from this region. Asteroid impact debris from faraway locations had rained down from the lunar sky, and it dominated the rock sampling all around Apollo 16. Young and Duke seemed to take it as a personal failure, until Mission Control reassured them, "Don't worry. We're going to have to put together a whole new picture of this place because of what you do see."

The two men drove more than 16 miles in the course of three EVAs. Their takeoff and redocking with *Casper* was successful, although various glitches kept everyone alert. As *Casper* headed back to Earth, Mattingly crawled out of the hatch to retrieve film canisters from the scientific instrument module (SIM), the suite of lunar surveying cameras and sensors in the service module at *Casper*'s rear.

Splashdown in the Pacific was on April 27, 1972. The aircraft carrier USS *Ticonderoga* took *Casper* to the North Island Naval Air Station at San Diego in California, where it unleashed one last, nasty surprise. Some of the harness lines on Apollo 15's parachutes had been slightly scorched by corrosive fuels dumped from the thruster system, so Apollo 16 was ordered not to vent any leftover fuel during reentry. On May 7, the highly reactive residues were drawn off into tanks on a wheeled cart. One of its tanks exploded, knocking out hangar windows, exposing dozens of technicians to toxic fumes and fracturing one man's kneecap. Fortunately, no lasting damage was done to anyone.

" THE MOON WAS THE MOST SPECTACULARLY BEAUTIFUL DESERT YOU COULD EVER IMAGINE. UNSPOILT. UNTOUCHED. IT HAD A VIBRANCY ABOUT IT, AND THE CONTRAST BETWEEN IT AND THE BLACK SKY WAS SO VIVID, IT JUST CREATED THIS IMPRESSION OF EXCITEMENT AND WONDER. "

Apollo 16 LM pilot Charles Duke, 2007

Young unpacks the LRV at the end of Apollo 16's third and final period of surface operations.

Apollo 16's LM approaches the CM, backdropped by the Sea of Fertility. Notice the crumpled panels, damaged by back blast from the LM's rocket engine at the moment of separation from the Descent Stage.

Apollo 16's CM orbits around the far side of the moon. The "up" or "down" orientation of the spacecraft was adjusted frequently to point its science instruments at specific targets.

A very crisp mapping photo taken by Apollo 16 helps explode the myth of a so-called "dark side of the moon." This hemisphere gets just as many hours of sunlight as the other.

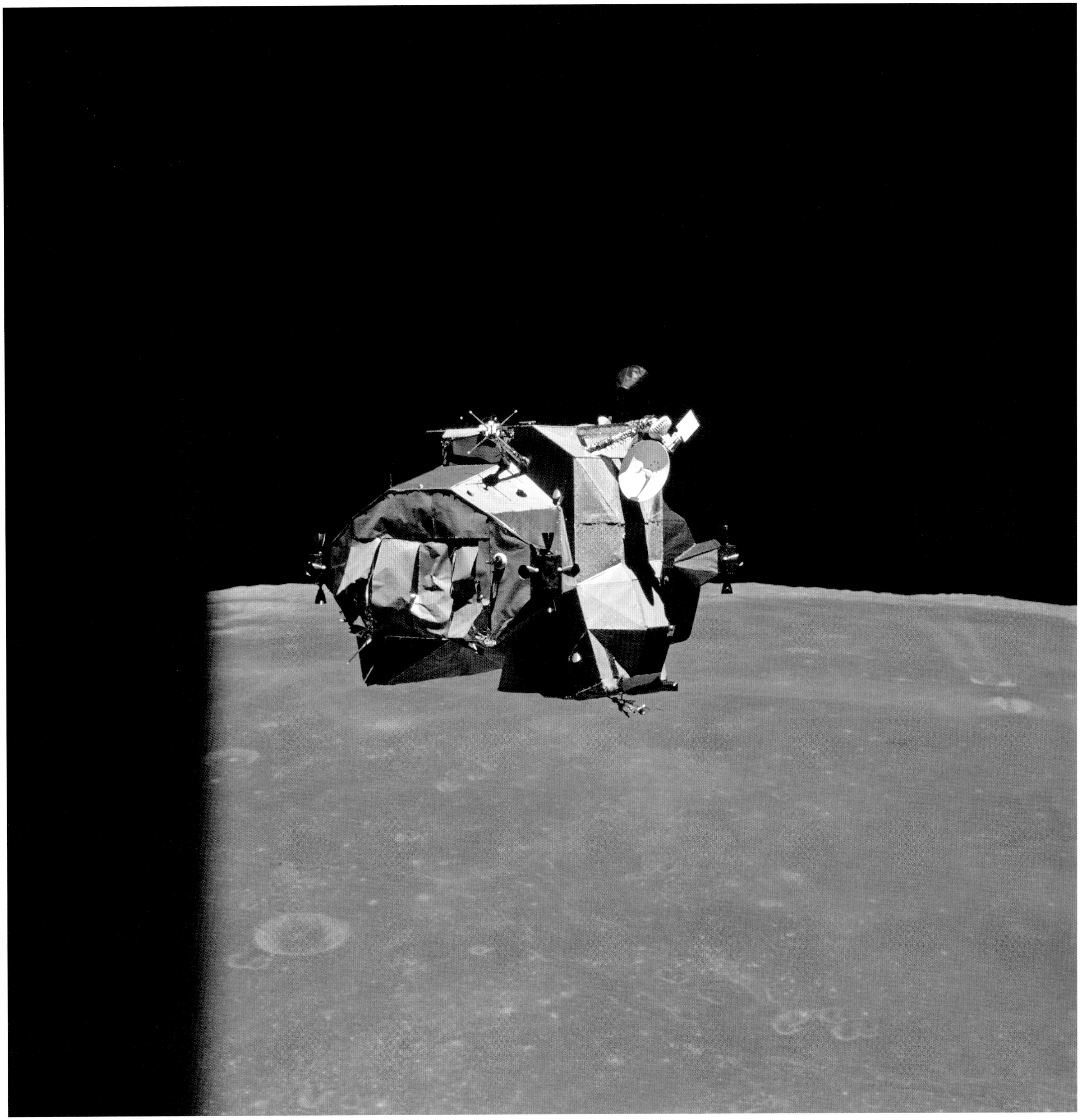

Scientist-astronaut Harrison Schmitt stands next to a huge, split lunar boulder during the third Apollo 17 EVA at the Taurus-Littrow landing site.

APOLLO 17 TAURUS-LITTROW

The last of the lunar landing missions was famous for including among its crew the first scientist launched by NASA. Lunar module pilot Harrison Schmitt made full use of his PhD in geology before he learned how to fly jet planes so that he could be accepted into the astronaut team. His Apollo colleagues began their space careers much more interested in rockets than rocks, but as their preparations for the moon got underway, all of them fell under the spell of the dramatic narratives that boulders, mountains, craters, and rilles could reveal, if only they learned to recognize the clues.

Apollo 20 was taken off the schedule five months after Neil Armstrong and Buzz Aldrin had "won" the moon race. Apollos 18 and 19 were canceled soon afterward, as budget cuts, and changing political priorities, impacted NASA. When Schmitt and his colleagues, Apollo 17 commander Gene Cernan and commander module pilot Ron Evans prepared for a dramatic nighttime launch on December 7, 1972, they knew that it would be a good few years before anyone followed them to the moon, but they never imagined that another half a century would pass, and still no one would have built upon their achievements.

Perhaps this explains why the photographic record from Apollo 17 is unusually rich. Everyone was determined to get as much out of this mission as possible. More Hasselblad magazines were carried than in previous flights, and the LRV roamed for a total of 22 miles, covering as much terrain as possible before the astronauts left the moon for the last time.

One of many highlights of the mission was the discovery of something colorful on the moon, while the astronauts were exploring a feature called Shorty Crater. Unlike countless other craters, this one seemed to be the product of volcanic activity. Just under the usual gray-white powdery topsoil, a very fine orange material was exposed by the sampling tools. The soil contained tiny glass beads, products of volcanism. The orange hues looked rusty, and in a sense, they were. It seemed as if iron interacted with water vapor during the ancient eruptions and turned into iron oxides. Not for the first time in the Apollo saga, water trapped and preserved inside the mineral structures upset what we thought we knew about the moon's history.

The End of Our Fascination?

Scientists were thrilled by such discoveries. The public, less so. NASA is so often criticized for its technical jargon, and for stripping language of its poetry and human impact. For instance, the noted American journalist and novelist Norman Mailer, dispatched by *Life* magazine to cover Apollo 11, was impressed by the professionalism of the mission controllers, and by their calmness and cooperative spirit, yet he complained that they "spoke in a language not fit for a computer of events that might yet dislocate eternity." Men had ventured to the moon, but somehow the true majesty of that fact was not reflected by NASA's dry presentation. The dangers of the mission—precisely those dangers that might have excited and moved the general public into understanding just how incredible this achievement really was—were somehow lost in translation, and perhaps even deliberately downplayed. As a consequence, the public became bored by lunar voyages, and it has never quite regained its initial fascination for the space adventure.

Hundreds of thousands of people are still fascinated by space exploration, and millions more take at least a passing interest, but those multitudes are perhaps not sufficient to ensure a long-term continuation of the human space program at a major national level. The drama of human rocket flight is as great now as it ever was, but astronauts do not seem to captivate entire societies they way they once could, and rockets have lost some of their ability to inspire awe. Matters are not helped by the fact that the International Space Station is confined to Earth orbit, and its day-to-day schedule of maintenance and docking activities does not

Apollo 17's LM, the last of the series, on the lunar surface. How long will it be before human explorers land there again?

Gene Cernan, last man on the moon, posing by the U.S. flag. The LRV's communications antenna is visible on the right.

Cernan takes the LRV for a test drive, staying within easy range of the LM at first, in case of any problems.

Panoramic assembly from multiple frames, with East Massif in center background. The scattered rocks are a sign of the moon's violent history over the aeons.

Cernan grips a corner of the flag, backdropped by a distant Earth. Apollo lunar mission commanders wore red stripes on their visor assemblies.

Harrison Schmitt qualified as a geologist before joining NASA. Here he uses a rake sampler to collect rock fragments too small to be picked up by hand.

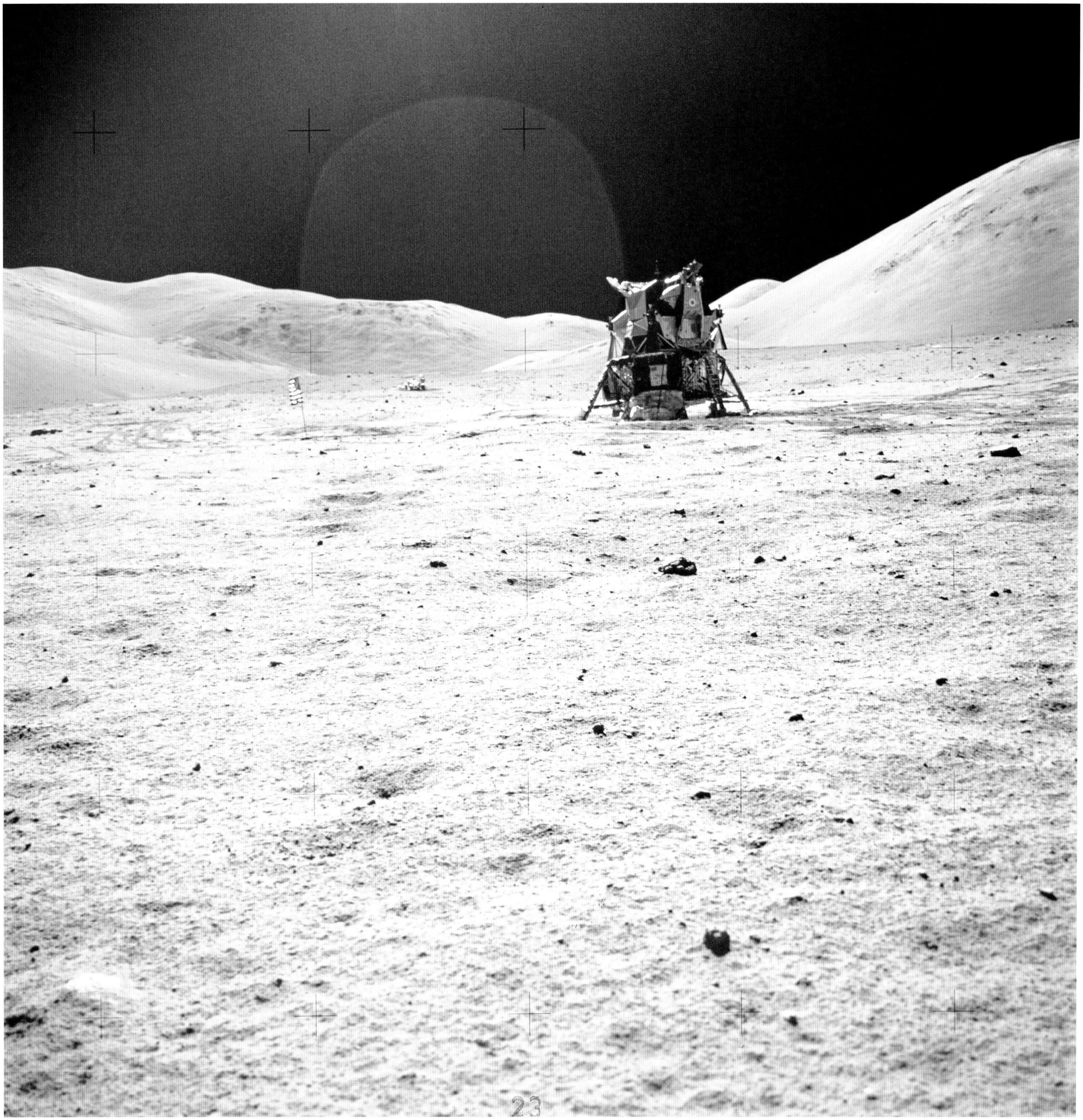

excite the public, trained by science fiction to expect that spaceships should actually go somewhere and explore something. There is a problem in the culture of space if young people no longer find as much excitement in the adventure as their parents did fifty years ago, when they, in their turn, were young. Perhaps NASA and its international partners need to speak to this generation in a more poetic way that engages our emotions as well as our minds. In recent years, many astronauts have tried to address this problem of public engagement, and some have proved adept at writing and lecturing for highly engaged audiences.

On the other hand, space exploration is enjoying a resurgence of public interest now that we are living in an age where, once again, it seems unusually difficult for humans to venture further than Earth orbit. Politically and culturally, there seems to be some appetite for pressing onward into the solar system, and for landing humans on Mars. There is a conception, also, that we should not bother going back to the moon because we have already done that. However, the moon is just three days' flight time away, and it would be a relatively safe place to practice the art of living on a nearby dusty alien training ground before we attempt the tougher and more hazardous business of surviving for months at a time on Mars, far from any hope of rescue.

We are not ready for the Red Planet just yet, so perhaps it is time to build on the legacy of Apollo and imprint just a few more bootprints in the lunar soil before journeying onward. Anyway, the moon itself still has many more secrets to give up: secrets that will tell us more about the one planetary body that preoccupies us more than any other. This one.

> "THE EVIL OF GOOD IS BETTER. WE'RE GOING TO TRY AND MAKE THINGS BETTER AND BETTER AND BETTER, MORE COMPLICATED, MORE EXPENSIVE, MORE TIME-CONSUMING, AND NOT ANY SAFER. THE KEY IS TO GET THERE SAFELY AND ADEQUATELY."

Apollo 17 commander Gene Cernan, December 2007

MOONSHOTS / MAKING TRACKS

175

> **WHEN YOU SHUT THE ENGINE DOWN-BOOM! AND YOU HIT. NOT REAL HARD, BUT WITH A THUMP. THAT'S WHERE YOU EXPERIENCE THE QUIETEST MOMENT A HUMAN BEING CAN HAVE. THERE'S NO VIBRATION. THERE'S NO NOISE. THE DUST IS GONE. ALL OF A SUDDEN YOU HAVE JUST LANDED ON ANOTHER WORLD SOMEWHERE OUT IN THE UNIVERSE, AND WHAT YOU ARE SEEING IS BEING SEEN BY HUMAN EYES FOR THE FIRST TIME.**
>
> Apollo 17 commander Gene Cernan, December 2007

These composite images show how individual film frames can be overlapped to create wider scenes. Modern software allows minor perspective misalignments to be corrected, but care has to be taken not to compromise the factual reliability of the images.

The rear right wheel fender of the LRV patched up with lunar maps, and clamps scavenged from an optical alignment telescope. The repair was needed after a rock hammer got underneath the fender and tore some of it away.

These photos of Cernan and Schmitt working alongside the LRV show how dirty their lunar suits became. Keeping dust out of spacecraft and living quarters will be a major design issue for future missions.

Shadows on the moon are certainly not always pitch black. Notice how how sunlight reflected from the light surface softened the shadows of these boulders, allowing plenty of detail to be seen.

We think of the moon as gray, but analysis of sliced and polished rock samples back on Earth reveals the rich and complex story hidden underneath the surface dust.

> I HAVE BEEN ASKED–OH, SO MANY TIMES OVER THE LAST YEARS– 'HOW DOES IT FEEL TO BE THE END? HOW DOES IT FEEL TO BE THE TAIL OF THE DOG? THE LAST ONE OVER THE FENCE?' I GOT ON MY SOAPBOX AS SOON AS WE CAME BACK. I SAID, 'APOLLO 17 IS NOT THE END. IT'S JUST THE BEGINNING OF A NEW ERA IN THE HISTORY OF MANKIND.'

I ALSO SAID, 'WE'RE NOT ONLY GOING TO GO BACK TO THE MOON, WE WILL BE ON OUR WAY TO MARS BY THE TURN OF THE CENTURY.' I SAID THAT IN 1973. AS YOU CAN WELL IMAGINE, MY GLASS WAS HALF-EMPTY FOR THREE DECADES. FORTUNATELY, NOW IT'S HALF-FULL AGAIN, AS LONG AS WE CONTINUE TO MOVE FORWARD.

Apollo 17 commander Gene Cernan, December 2007

MOONSHOTS / MAKING TRACKS

183

"WHEN OUR MISSION ACTUALLY FLEW, THERE WERE FAR FEWER SYSTEM FAILURES. EVERYTHING JUST WORKED BEAUTIFULLY FROM A HARDWARE AND A SOFTWARE POINT OF VIEW, WHICH WOULDN'T HAVE HAPPENED IF PEOPLE HAD BEEN LETTING US DOWN ANYWHERE ALONG THE LINE."

Apollo 17 LM pilot Harrison Schmitt, 2000

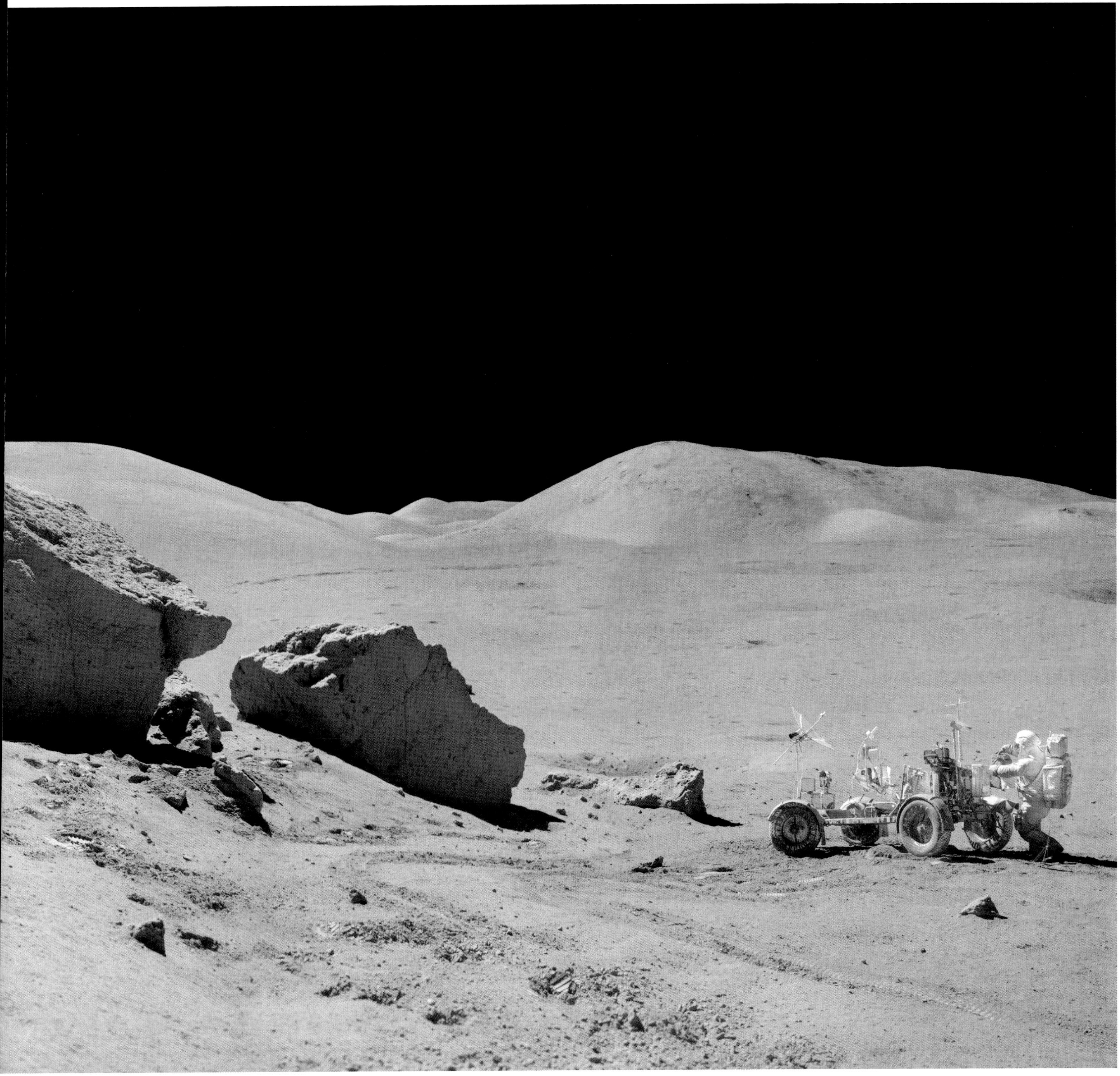

The final shot of a human walking on the moon (for now, at least). This is Schmitt, photographed by Cernan shortly before both men climbed back into the LM for the last time.

Gene Cernan in the LM at the end of surface operations. He looks tired and dirty, but happy.

The Scientific Instrument Module (SIM) bay in the Service Module behind Apollo 17's CM. As the spacecraft coasted back to Earth, CM pilot Ron Evans made a space walk to retrieve the exposed film rolls.

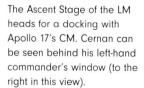
The Ascent Stage of the LM heads for a docking with Apollo 17's CM. Cernan can be seen behind his left-hand commander's window (to the right in this view).

An unusual close-up of the CM's right-side rendezvous window. The thin, metallic strips of thermal and radiation shielding burn off during reentry.

A famous view of Earth from Apollo 17. This was the first time that an Apollo trajectory made it possible to photograph the south polar ice cap. Almost all of Africa is visible.

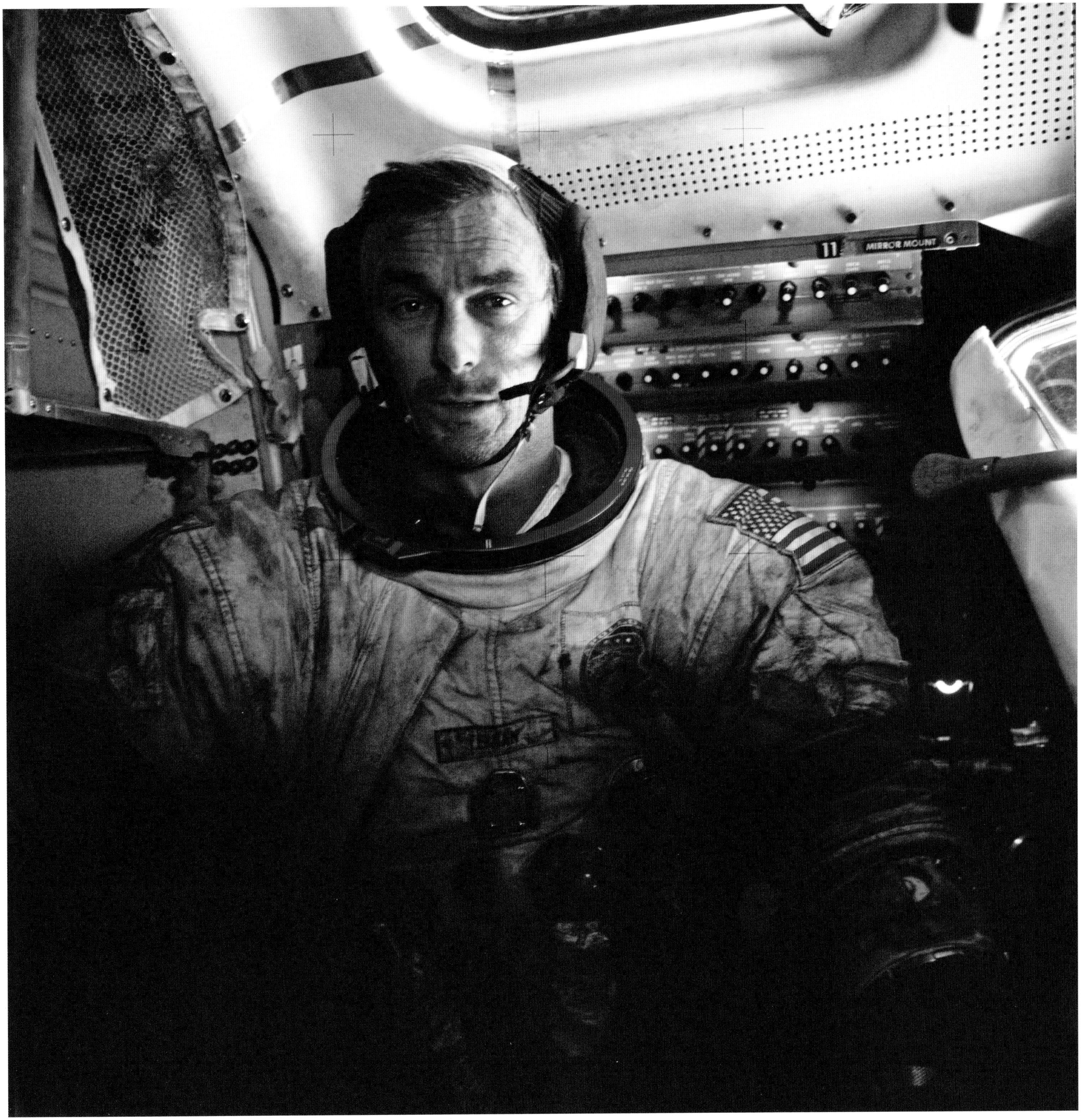
11
MIRROR MOUNT

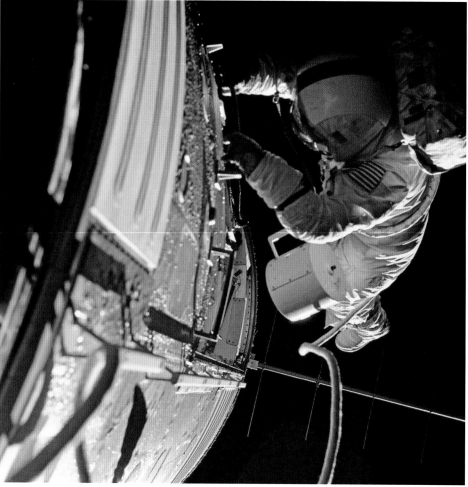

MOONSHOTS / MAKING TRACKS

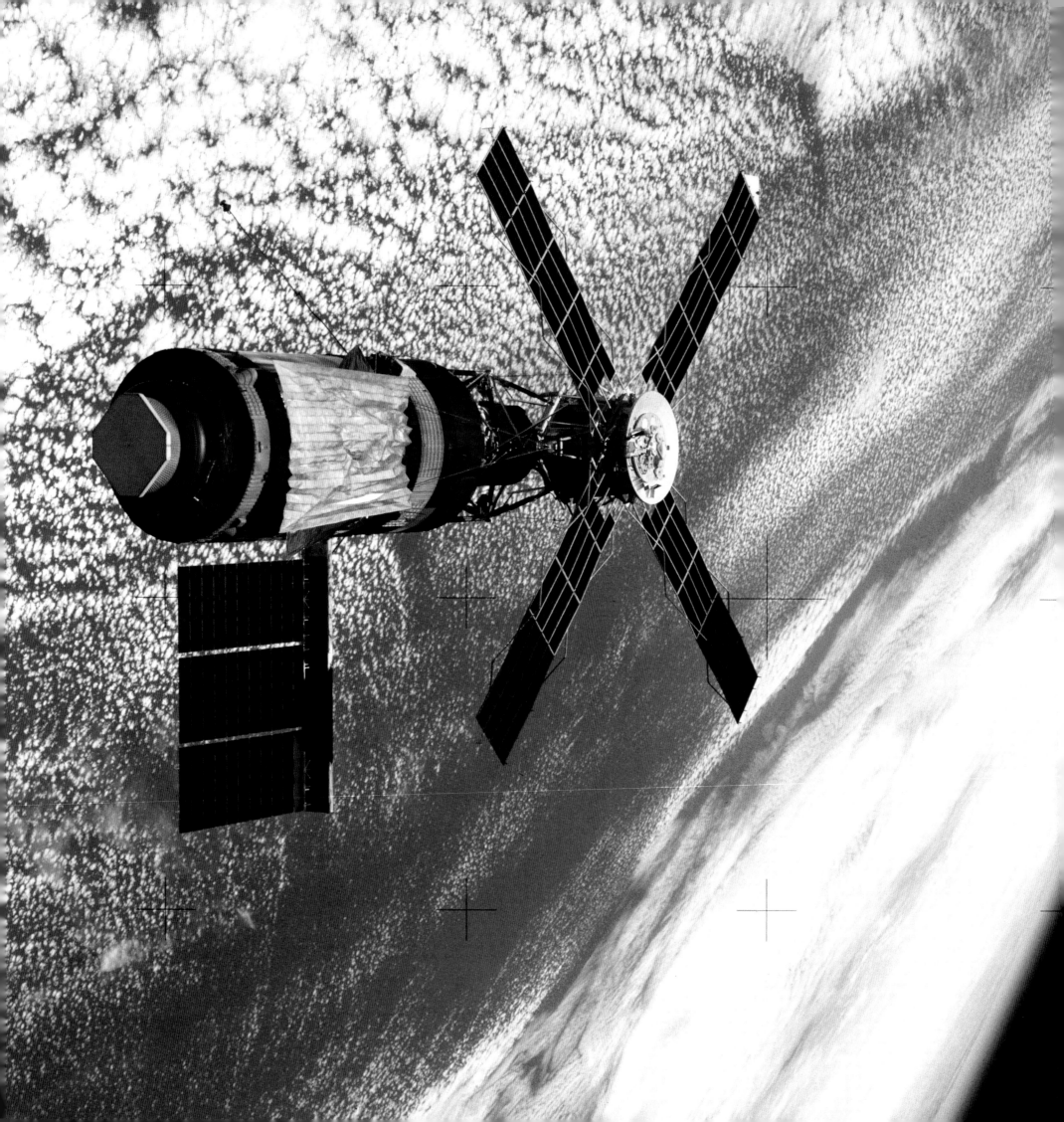

Skylab in orbit, seen here after emergency repairs. In the absence of the original thermal shielding (which was torn off during launch) a Mylar sheet shields the main compartment from the sun's heat.

5 LIVING IN ORBIT

Missions to the moon took just a few days to complete. In the mid-1970s American astronauts learned how to stay in space for much longer periods.

The Apollo system was designed to accomplish more than just reaching the moon. NASA wanted to use the immense lifting capacity of the Saturn V to assemble a huge modular space station in Earth orbit. Budget cuts forced planners to scale down their ideas, but one project survived. The Skylab Orbital Workshop, launched in 1973, still holds the record as the largest spacecraft ever sent aloft on a single rocket. It gave American astronauts their first taste of long-duration orbital flight.

A major factor in Skylab's favor was its economical design. It was already half-built even before Congress approved it in 1969. By then, Apollo lunar landing missions 18, 19, and 20 had been canceled, and one unflown Saturn V was available for conversion into a space station launcher, along with three Apollo command modules to serve as crew ferries (and a fourth kept spare as a potential rescue vehicle). The third, uppermost stage of a Saturn V, was adapted as Skylab's central component. Its huge, empty fuel tanks were converted into living quarters in just thirty-six months. Skylab lifted off on May 14, 1973, on the last Saturn V ever to fly.

The launch seemed flawless, but it turned out that thermal shielding on the lab's outer skin had been torn off by aerodynamic stresses. One of the solar panels was jammed shut, and the second had torn away, like a bird's wing ripped off at the shoulder. Skylab's electrical systems were dead without them. There was no choice but to postpone the launch of the first three-man crew on a separate, smaller rocket (a Saturn 1B) while NASA assessed the damage. Astronauts Joe Kerwin and Paul Weitz, along with their commander Pete Conrad, hastily trained for an emergency spacewalk, testing out repairs in a giant water tank, using a mockup of the station.

On May 25, Conrad and his crew docked with Skylab and began their "Mr. Fix-it" spacewalk. They freed the jammed solar panel so that it could open up, then unfurled a metal foil sun shield to protect Skylab's main compartment from the heat of the sun, tying it down to convenient fixing points with thin straps. This was no job for beginners, but NASA had never tried anything like this before.

When the astronauts went inside Skylab, they found the interior was dangerously hot. They spent their first few days in the cramped docking module at the front of the station, until the sunshade had taken effect, and they could bring Skylab's systems to life. Nothing could dampen the mood. Conrad and his crew had saved the huge spacecraft from disaster in a dramatic rescue that made headline news around the world.

Two other crews visited Skylab over the subsequent year. As well as conducting medical experiments, they operated a solar observatory with X-ray, infrared, visible light cameras. Skylab was a tremendous success, although one of its three-man teams came close to mutiny, and changed forever the relationship between managers on the ground and astronauts in orbit.

Ed Gibson, Bill Pogue, and commander Jerry Carr were the final Skylab crew, and they had something to say about the interior design. "We really need a better sense of up and down with a proper difference between the floor and ceiling," Pogue commented on the radio link with Mission Control. "The layout in the docking adapter is so lousy I don't even want to talk about it." Engineers had used floors, ceilings, and walls to mount control panels and storage lockers, so that every available area could be put to use. The effect was disorienting, to say the least. Light fittings and workstations were sometimes on the "ceiling," and sometimes on the "walls." The astronauts hardly knew from one moment to the next which way round they were.

Next on their list of grievances was the station's decor. "The color scheme in here has been designed with no imagination," Gibson moaned. "All we've got is about two shades of brown, and that's it for the whole lousy spacecraft interior." The crew's fireproof garments were as drab as the walls, and the artificial fabrics were stiff and prickly against the skin. They all yearned for floppy T-shirts and casual, comfortable gear in cheerful colors.

There was more to worry about than bad fashion. Skylab's bathroom was unpleasant to use. The metal floor was designed as a hygienic wipeable surface; it had no firm footholds, and at wash time, the astronauts slithered like sardines in a can. Having a good wash took up not minutes but hours, because every last droplet of water that strayed outside the plastic shower enclosure had to be mopped up to prevent moisture seeping into electrical equipment and short-circuiting the works.

All this unexpected housework cut into the few precious moments set aside for the astronauts to relax. When Mission Control tried to discuss this problem, Carr was too wound up to listen. "Off-duty activities? You gotta be kidding! There's no such thing up here. On a day off, the only difference is, you have time to clean the shower!"

This was the real problem. The crew's tight work schedule stopped them from enjoying the actual experience of being in space. Every spare moment, they huddled around Skylab's single small window, staring in childlike fascination at their home world, seeing it as very few people before or since have ever had a chance to see it. But there weren't many spare moments.

Carr finally called a strike. "I get the feeling that there's some hassle about who on the ground gets our time, and how much of it. We hoped you would have got the message that we did not plan to operate at this pace. We need time for relaxation." For an entire day, the three men did nothing but stare out of the window, take photographs, or catch up on personal tasks. Meanwhile, NASA flight directors had a rethink and decided, to nobody else's great surprise, that maybe they were pushing the astronauts too hard.

Long-term orbital life was a new experience for everybody at NASA, and the occasional complaints were seen as valuable learning tools, both then and now. Throughout the design process for today's International

The last Saturn V lifts off, carrying Skylab, NASA's first space station.

Artwork showing the generous living and working space inside the Skylab space station.

Space Station (ISS), Carr and Pogue served as special consultants to NASA. In advanced middle age, they were still trim and fit enough to clamber into spacesuits and practice station procedures in that familiar giant water tank. After more than a quarter of a century, the difficulties of life on Skylab remained fresh in their minds. They claimed a special field as their own: how not to design a space station.

The astronauts who live and work aboard ISS are consulted about everything, from spacewalk procedures to the choice of beverages. But no matter how many astronauts offer their opinions, a lot comes down to personal taste. In 1991, someone decided that pink was a restful color suitable for the ISS living quarters. Shuttle pilot Joe Allen wasn't so sure. "I don't live in a pink house. Why would I want to live in a pink space station?" he said.

Saturn's Sunset

On completion of Apollo and its associated programs, there were no more payloads on the horizon large enough to justify the launch of any more Saturn V rockets. The image on page 199 shows the last operational Saturn V lifting off on May 14, 1973, carrying the Skylab orbital workshop, America's first space station. Historians and economists can argue for years about how we managed so quickly to abandon a rocket of such proven power and reliability, but at the time it was felt that the huge ground facilities and thousands of staff required to fly the Saturn were no longer justifiable.

Likewise, the smaller Saturn IB rocket was retired. The last one, launched on July 15, 1975, carrying Tom Stafford, Deke Slayton, and Vance Brand, was headed for the first international crewed space mission, a peaceful orbital hookup with a Soviet Soyuz spacecraft. The Space Race was at an end, although the Cold War missile standoff on the ground would not be resolved until the Soviet Union's sudden collapse in 1989.

Gerald Carr, commander of the third and final crew to visit Skylab (in May 1973) jokingly pretends to balance William Pogue on his fingertip.

Portraits of a wounded bird. The absence of Skylab's left side solar array is obvious. The Mylar sheet on the top of the living quarters, deployed during an emergency repair, helped stabilize the temperature.

Owen Garriott works outside Skylab, a spacecraft that still holds the record as the largest launched on a single rocket.

" BY 1973, WE HAD BEEN TO THE MOON. WE HAD OUR FIRST SPACE STATION, AND WE HAD MULTIPLE PROBES GOING UP TO THE PLANETS. ALL THIS WONDERFUL STUFF HAPPENED IN THE SPAN OF JUST FIFTEEN YEARS. "

Private spacecraft pioneer Burt Rutan, 2004

An early morning view of Pad B, Launch Complex 39 in July 1975, showing the last Saturn 1B, and the last Apollo CM, preparing for a mission to dock for the first time with a Russian spacecraft.

A painting by the noted aerospace artist Robert McCall shows the diplomatic joining of a Russian Soyuz with an Apollo spacecraft, signifying an end to the competitive superpower Space Race.

OLD ENEMIES UNITE APOLLO AND SOYUZ

When President John F. Kennedy urged America to land a man on the moon in that famous speech of 1961, he did so in an age when national self-confidence was at its height. The economy was in good shape. New technologies were transforming the industrial sector. Energy was cheap, employment levels were high, and the Soviet Union seemed a palpable threat to all this. By the turn of the decade, the national mood was very different. Civil rights disturbances, student riots, and the toxic debate about the Vietnam War had polarized the country. While NASA's achievements were widely celebrated, its key job was regarded as essentially done. It had beaten Russia to the moon, and public support for expensive space projects had diminished.

In 1972, President Richard Nixon and Russian Premier Alexei Kosygin signed a deal that allowed Russian and American space engineers to visit each other's countries and plan a joint mission, so that both sides could rein in some of the huge costs of competition in space. This was a diplomatic breakthrough, designed in part to try and soothe East-West geopolitical tensions.

Russia's Soyuz capsules were being prepared to dock with the new Salyut space station, the first of an expanding series.

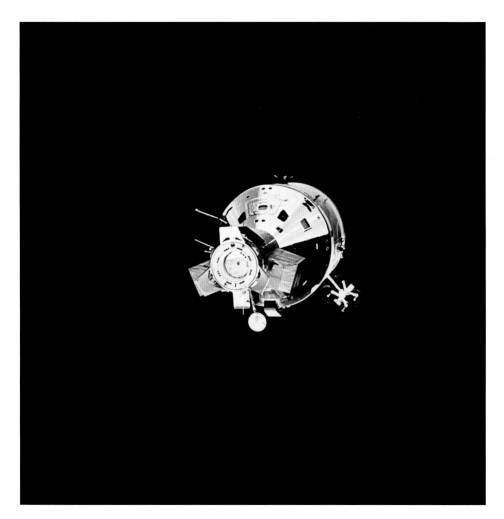

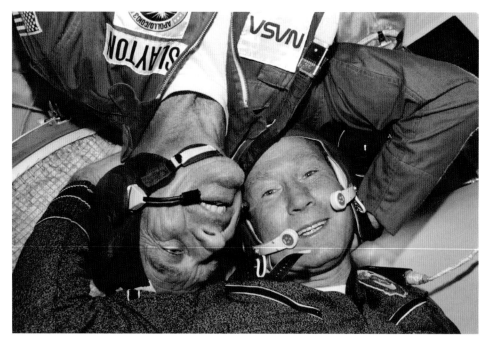

Scenes from a reconciliation: The two spacecraft take a good look at each other, and NASA astronaut Deke Slayton poses with Russian commander Alexei Leonov shortly after docking.

Meanwhile, apart from the hardware allocated for Skylab, Apollo was almost expended. Manufacturing of components had ceased, and there was just one final mission on Apollo's books. In terms of scientific knowledge, or advancing the exploration of space, it would be almost entirely without value. Politically, it was expected to yield huge benefits. It was called the Apollo Soyuz Test Project (ASTP).

ASTP's planners knew that persuading engineers from rival superpowers to work toward the same engineering goals would be an achievement in itself. Apollo and Soyuz used incompatible docking systems, and the crew compartments worked at different atmospheric pressures. The solution was a small, jointly designed adapter that attached to the nose of the Apollo, in exactly the place where a lunar module normally would have docked. The "front" end of the adapter had a probe capable of mating with the nose of a Soyuz. The adapter would serve as a buffer between the two ships' atmospheres, slowly equalizing the pressures between the docked craft.

On July 17, 1975, the docking took place. The two crews shook hands, clambered into each other's ships, and shared food and drink. Genuinely moved, Russians on the streets of Moscow gave great bear hugs to foreign visitors. It seemed a breakthrough for peace, yet it would take twenty years for Russian and American space technology to get that close again. The new political détente proved fragile, until the collapse of the Soviet Union in 1989 changed the relationships once again.

Gemini and Apollo veteran Tom Stafford commanded the NASA crew. His two crewmates were Donald "Deke" Slayton and Vance Brand. The Soyuz commander was another veteran flier, Alexei Leonov (the first man to walk in space in 1965). Valeri Kubasov was his copilot.

Politics faded into the background as close friendships were formed between the supposedly rival space agencies. One story illustrates the atmosphere between them. When NASA's smartest guidance people expressed concern that the Soyuz software was not equipped to deal with the countless pages of orbital math required for the hookup, their equally brilliant Russian counterparts handed over a single sheet showing how the math could be simplified. Respect between the two teams was swiftly established. As Vance Brand recalled in 2000, "We thought the Russians must be pretty aggressive people, and they probably thought we were monsters. We quickly broke through that, because when you deal with people in the same line of work as yours, you discover that they're human beings."

After ASTP, America's astronauts were grounded for six years, waiting for a new space vehicle to become operational. Meanwhile, Soyuz ships continued to launch on a regular basis as the Russian space station program continued its steady advance.

An Apollo-Soyuz view of clouds reshaped by contrails from jet liners. Astronauts witness many such changes from orbit, and we would be wise to listen when they tell us what they've seen.

THE SHUTTLE ERA

Even as Apollo 11 made its epic touchdown in July 1969, NASA knew that attitudes toward space had changed. Public appetite for rocket adventures had abated, and neither the White House nor Congress was keen to promote huge budgets for space adventures. Soviet space achievements no longer scared anyone. The global economy was on a long downturn, and increased energy costs were hitting everyone's pockets. Bitter arguments about the Vietnam War had soured many people's enthusiasm for noisy nationalistic expressions of technological supremacy.

It took a while for NASA to acknowledge that its future would have to be leaner and less glamorous than in the days of Apollo, but the truth was that the gigantic Saturn V launch vehicles had become unacceptably expensive to fly, because they were thrown away after single use. In response to this early curtailment of Saturn's career, President Nixon was asked to support development of a reusable winged space shuttle that would piggyback on a separate and similarly reusable winged booster. The shuttle would carry astronauts, along with components for a permanent space station.

Space Shuttle Columbia stands on the pad, prior to the inaugural flight of NASA's new multipurpose winged vehicle on April 1981.

Now that Apollo had accomplished its principal task, Nixon did not expect much public or congressional support for another huge NASA scheme. The station was put on indefinite hold, and NASA salvaged its shuttle concept essentially by promising that it would pay its way by carrying commercial satellites into orbit, turning the once-proud space agency into a trucking company. In a short but key memo, Secretary of Defense Caspar Weinberger advised Nixon, "We don't want the world to think our best years in space are behind us. NASA's proposals have some merit. We keep slashing their budget because we can, and not necessarily because they are doing a bad job."

NASA was given the go-ahead, but only so long as the fully reusable carrier stage was lost in favor of cheaper solid rocket boosters that could be partially refurbished after each flight and a huge liquid fuel tank that was completely jettisoned. The space shuttle started life as a compromised design. New technology had to be developed on a shoestring, despite the incredible challenges that had to be solved. For instance, the shuttle's three main engines had to last many flights, at a time when no one had figured out how stop an engine melting after just one use. The heat shield tiles were another nightmare, requiring thirty-four thousand ceramic panels to be glued, one at a time, onto the winged Shuttle's exterior.

At last, on April 12, 1981, shuttle *Columbia* was launched, and its success seemed to banish all the difficulties. Newly elected president Ronald Reagan was sympathetic to the romance of America's return to astronaut flights after a layoff of six years. Sister shuttles *Challenger*, *Discovery*, and *Atlantis* entered service over the next few years, and all four vehicles achieved astonishing successes during what appeared to be an increasingly routine schedule. In January 1984, Reagan announced that NASA should build a new space station in partnership with international allies. At last, the shuttle fleet could fulfill its ultimate purpose as a delivery system for a giant orbiting outpost. NASA celebrated its flagship spacecraft as "the most complex machine in history." In those innocent first five years of orbital triumphs, few people outside the space industry realized that this fabulous complexity was not a virtue but a hazard. Judging by surface appearances, the shuttle was a science fiction dream come true, an awesome winged spaceship flying time and again into orbit and coming home to land like a plane. But nothing was more certain to turn dreams into nightmares than America's confidence that a machine powered by five furious infernos of hydrogen, oxygen, aluminum powder, and ammonium perchlorate could be considered so tame that its launches were barely worth remarking upon anymore.

Nature Cannot Be Fooled

On the wintry late morning of January 28, 1986, STS-51L shuttle *Challenger* was launched from a pad gnarled with icicles and ascended into an uncharacteristically frigid Florida sky. The right-side solid rocket booster sprang a leak immediately on ignition. Freezing weather had compromised the rubber seals between its cylindrical segments. A jet of flame scorched the side of the huge liquid fuel tank to which the boosters and *Challenger* itself were attached.

Seventy-three seconds after liftoff, commander Dick Scobee got the all clear from Mission Control to throttle up *Challenger*'s trio of liquid-fueled engines for the last push toward orbit. That was the moment when the wayward flame finally penetrated the skin of the external fuel tank, instantly igniting the potent fuels within. The tank exploded and STS-51L disintegrated at twice the speed of sound. On board were commander Dick Scobee, pilot Mike Smith, fellow astronauts Greg Jarvis, Ron McNair, Ellison Onizuka, and Judy Resnik, and one non-astronaut civilian, Christa McAuliffe, participating in a unique program aiming to send a high school teacher into space.

Renowned physicist Richard Feynman was invited to join the subsequent board of inquiry, the Rogers Commission. With the TV cameras running, he dipped a piece of rubber into a glass of iced water and showed how it hardened when cold. "Do you suppose this might have some relevance to our problem?" he asked, knowing very well that it did. He had created a vivid demonstration of the notorious O-ring problem that led to *Challenger*'s destruction. But Feynman wanted to probe further. "If NASA was slipshod about the leaking rubber seals on the solid rockets, what else would we find if we looked at the liquid-fueled engines and all the other parts that make up a shuttle?" he asked.

Feynman made unauthorized trips to NASA facilities where he could speak to junior engineers in private. He wrote later, "I had the definite impression that senior managers were allowing errors that the shuttle wasn't designed to cope with, while their engineers were screaming for help and being ignored." He had identified a serious cultural problem within NASA at that time. In the last days of the commission, Feynman made a plea for greater realism. "For a successful technology, reality must take precedence over public relations, because nature cannot be fooled."

Regaining Trust

The shuttle fleet was back in action by September 1988 with *Discovery* leading the return to flight status. Three years later, as the dust settled and NASA found its feet once again, a particular mission caught the public imagination. It was widely hailed as among the most worthwhile projects ever undertaken by astronauts.

Launched by shuttle *Discovery* in April 1990, the Hubble space telescope began its career in disarray. Its main mirror turned out to be faulty because of fundamental manufacturing errors. The costly telescope that was supposed to deliver our best views of the cosmos was essentially blind. Fortunately, Hubble was designed for servicing by shuttle astronauts. NASA devised an ingenious system of lenses and secondary mirrors that could refocus the telescope, and on December 2, 1993, shuttle *Endeavour* lifted off carrying a rescue crew.

Working as two separate duty teams, astronauts Story Musgrave, Jeffrey Hoffman, Thomas Akers, and Kathryn Thornton completed the repairs to Hubble in five back-to-back space walks totaling thirty-five hours. Hubble was pulled into *Endeavour*'s cargo bay using the Canadian-built remote manipulator robot arm, a versatile piece of

STS-1 lifts off on April 12, 1981, carrying Americans into space after a furlough of six years.

On a mission designated STS-41B (actually the tenth Shuttle flight) Bruce McCandless tests a Manned Maneuvering Unit (MMU) and tries out the foot restraints on the Shuttle's Canadian-built Remote Manipulator arm.

A chase plane captures an incredible view of the vapor trail at the start of the seventh Shuttle mission on June 18, 1983.

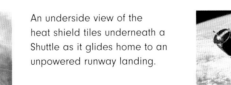

An underside view of the heat shield tiles underneath a Shuttle as it glides home to an unpowered runway landing.

Dale Gardner uses an MMU to approach the spinning Westar VI satellite during the November 1984 STS-51A mission, before retrieving the satellite for repair on Earth. It was successfully relaunched in 1990.

Sally Ride, the first American woman in space, during the STS-7 mission in June 1983.

STS-49 astronauts Kathryn Thornton (foreground) and Thomas Akers test space station assembly tools in May 1992.

After landing, a Shuttle is surrounded by support vehicles after touchdown. The flexible hoses remove any lingering traces of propellant in the maneuvering thruster tanks.

equipment that has featured in many shuttle flights. A self-contained package of corrective optics was eased into a side hatch on Hubble. The astronauts also installed new gyroscope control systems before gently easing Hubble free of the *Endeavour* and back into an independent orbit.

Hubble's mirror may have been flawed, but the spacecraft overall turned out to have been well designed. The access doors opened properly, and the new equipment slid into place according to plan. The spacewalks were spectacular, and NASA's regained confidence seemed infectious. Some political critics began to revise their views, perhaps unwilling to appear ungracious in the light of an obvious headline-grabbing success. The complex spacewalks also demonstrated that building a space station need not be as foolhardy a scheme as some commentators had imagined.

Hubble transformed our understanding of the universe, and became one of the best-loved scientific instruments ever created. It also proved that there is no simple choice in space exploration between human and machine operations. We achieve more when both elements are combined.

Back to Basics

On February 1, 2003, NASA repeated its earlier mistake of ignoring known flaws in the shuttle. Returning from a microgravity science expedition, shuttle orbiter *Columbia* disintegrated during reentry. Mission commander Rick Husband, his copilot William McCool, and colleagues Laurel Clark, Kalpana Chawla, Michael Anderson, and Ilan Ramon lost their lives.

All seven aboard were killed. A suitcase-sized chunk of foam insulation from the external fuel tank had peeled off during launch, slamming into the front of *Columbia*'s left wing and making a small but ultimately lethal hole in the heat-resistant panels. Minor insulation impacts had occurred in previous missions, but were ignored.

The *Columbia* Accident Investigation Board (CAIB) judged it safer for astronauts in the future to ride in self-contained capsules that can pull away from their launch vehicles in case of emergencies. The shuttles never had any such means of escape, so the CAIB urged that they should be retired as soon as possible. The U.S. president at that time, George W. Bush, agreed, and authorized NASA to build a multipurpose capsule, the *Orion*, launched on an expendable rocket.

In the next twist to the saga, President Barack Obama, faced by a global financial crisis, slashed these plans, leaving America reliant on rides aboard Russian Soyuz capsules for access to the International Space Station, and urging the development of small, privately developed spacecraft for accessing low Earth orbit. NASA's *Orion*, meanwhile, was to push into the depths of space, perhaps beyond the moon, and ultimately, toward Mars. The plans for further space exploration were vague then, and they remain hazy today. There is talk of astronauts reaching Mars by the 2030s.

This lack of certainty puts the space shuttle era in a new light. In the final analysis, it was a pretty capable machine, a true workhorse of space. It was as if a winged spaceplane from the cover of an old science fiction magazine had taken flight for real. The shuttle fleet's launch record of 133 missions safely completed out of 135 would be considered highly impressive for any other rocket family, so those two notorious disasters have to be seen in context; and the shuttles certainly achieved most of what was expected from them, carrying out tasks ranging from medical experiments to space probe deployments and the piece-by-piece construction of a huge space station. They launched 3 million pounds of cargo, and more than six hundred astronaut seats were occupied. As we pay tribute to the shuttles, we can only hope that something even more remarkable will replace them before too long.

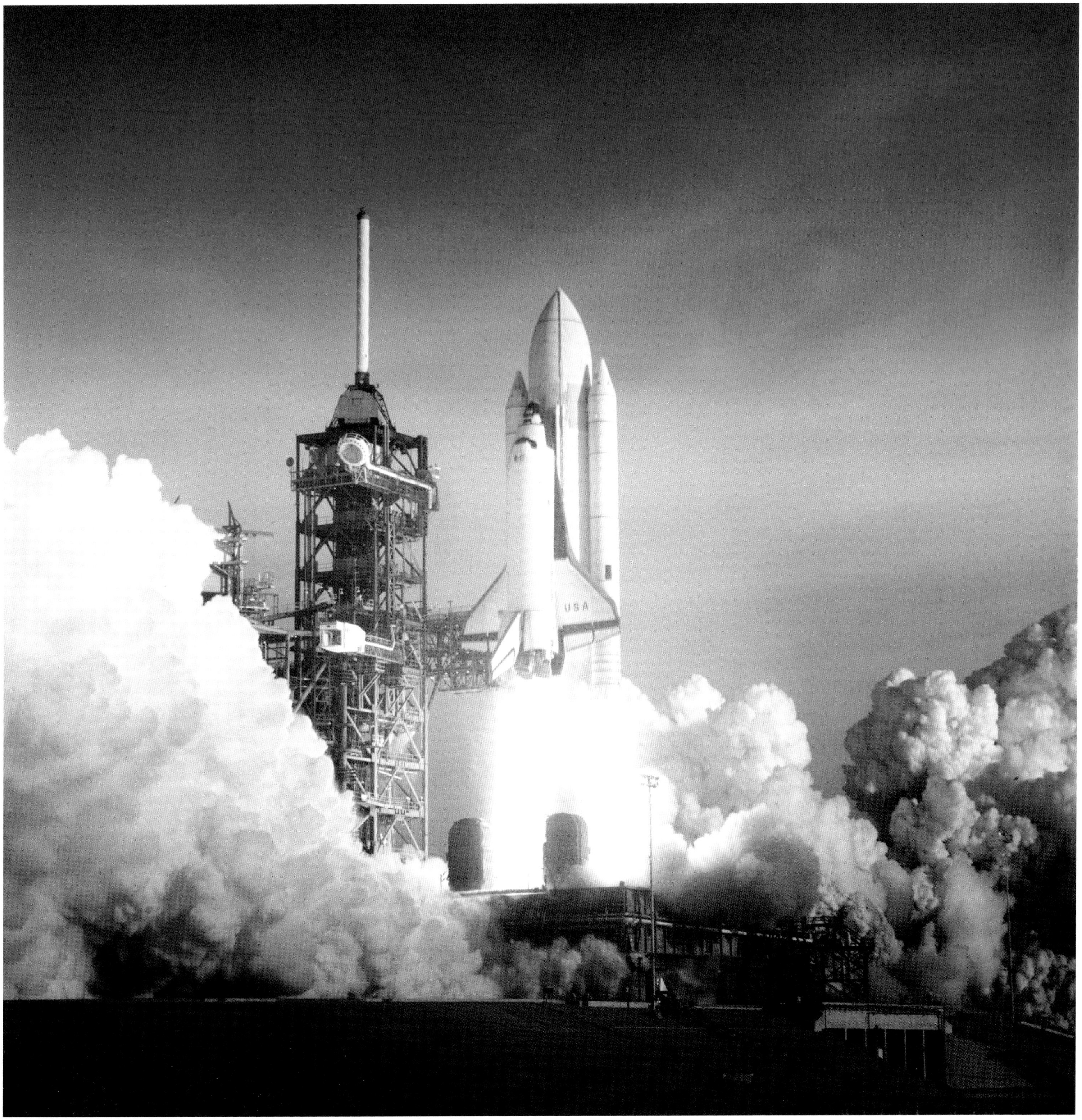
USA

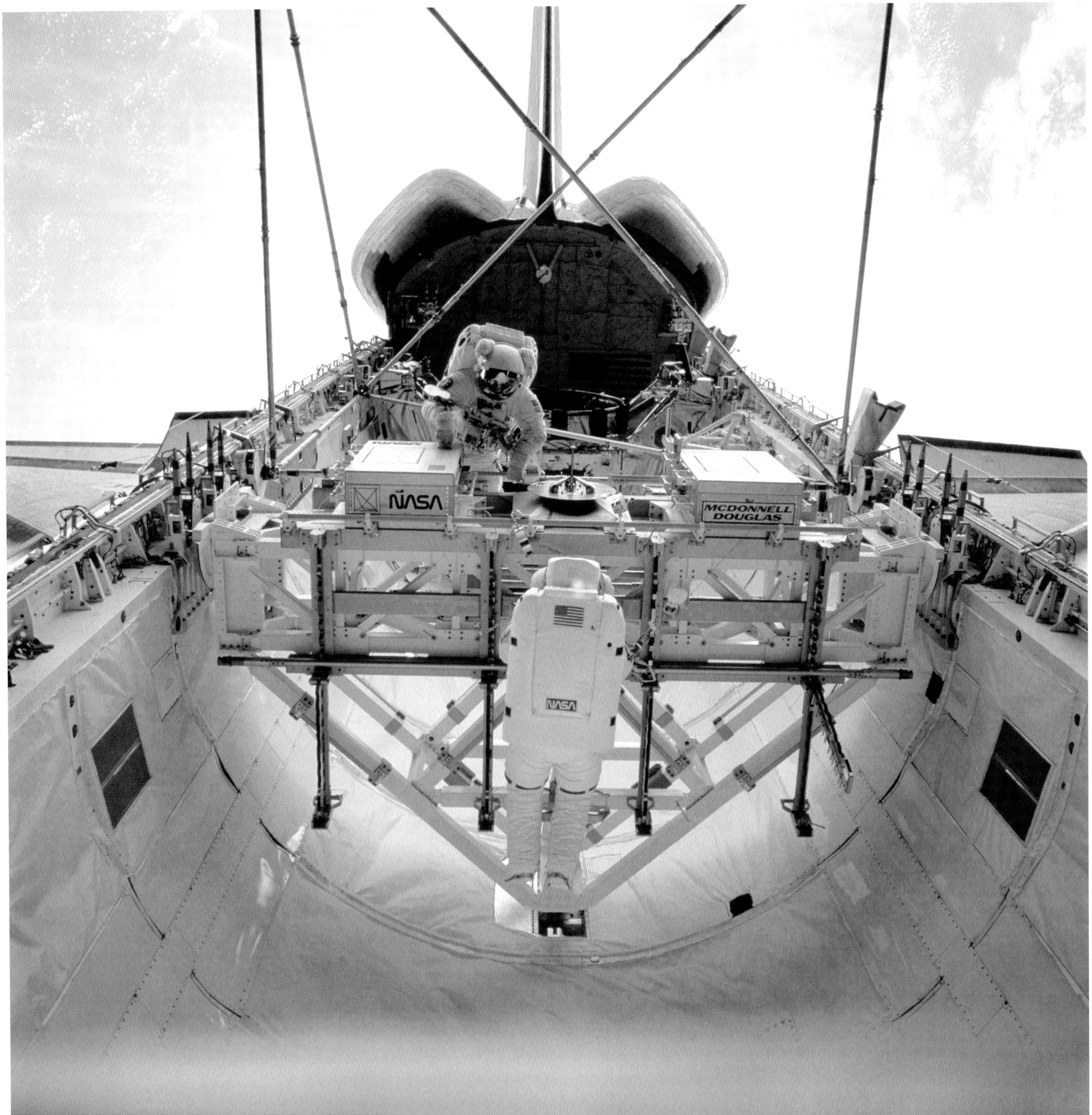
NASA
McDONNELL
DOUGLAS
NASA

Our dynamic home world, seen from orbit. The vast plume of Klyuchevskaya Sopka in the Russian Far East, a volcano that erupted in October 1994, and multiple thunderstorm cells developing southeast of Hawaii.

Shuttle missions in support of the Hubble Space Telescope proved to be among the most scientifically valuable and publically popular enterprises in NASA's astronaut program.

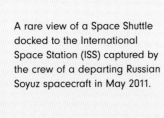

A rare view of a Space Shuttle docked to the International Space Station (ISS) captured by the crew of a departing Russian Soyuz spacecraft in May 2011.

AN ISLAND IN THE SKY

The common public perception is that the Space Race between Soviet Russia and the United States came to a sudden end with the lunar touchdown of Apollo 11 in July 1969. In fact, a degree of competitive tension persisted for another two decades, until the fall of the Berlin Wall and the end of the Soviet era. While NASA worked throughout the 1970s on development of the space shuttle, the Soviet Union answered with a steady expansion of its space station capability, starting with small Salyut orbiting platforms and culminating in the multi-module Mir complex.

In 1981, a new U.S. president, Ronald Reagan, was persuaded that NASA should embark upon its own station, and he introduced a unique element. In his State of the Union address of January 25, 1984, he announced, "Tonight, I am directing NASA to develop a permanently manned space station and to do it within a decade. NASA will invite other countries to participate so we can strengthen peace, build prosperity, and expand freedom for all who share our goals."

For some years afterward, NASA found itself stuck in political limbo, constantly redesigning the space station to meet changing political and scientific requirements. However, in the 1990s a seismic shift in history resulted in one last redesign of the station, and a new twist to its political justifications. Reagan's somewhat suggestive name for it, *Freedom*, was dropped, for there was no longer any suggestion that the Soviet Union might be a rival in orbit. The Soviet Union had ceased to exist...

In June 1992, NASA administrators met with their Russian counterparts to chart a new course. An era of cooperation would begin by launching shuttle missions to the Mir station, while the American-led station would now include Russian modules, as well as major elements from Europe, Japan, and Canada. The huge project finally became known as the International Space Station (ISS).

The motivations for the deals with Russia were complex. One factor was the need to protect Russia's rocket industry from economic turbulence while a market economy slowly

STS-101 Endeavour's view of the first core modules of ISS in May 2000, after attachment of NASA's Unity docking node to Russia's Zarya power module.

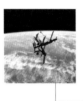
Mir, the last in a series of space stations constructed in the Soviet era. NASA shuttles and astronauts visited Mir repeatedly in the 1990s, establishing mutual trust ahead of further collaborations with Russia.

Russian cosmonaut Valery Polyakof keeps watch while the STS-63 Shuttle makes an exploratory close approach to Mir in February 1995.

A spectacular wide-angle view of STS-71 Shuttle *Atlantis* after NASA's first secure docking with Mir in June 1995.

NASA's Destiny science module is seen here after installation on ISS in 2001, as the sprawling structure continues to expand.

Living in space seems routine, but will we be able to maintain these costly projects? STS-113 astronauts John Herrington (left) and Michael Lopez-Alegria work outside ISS in November 2002.

began to emerge. Collaborative arrangements ensured that a wide range of highly capable Russian launch vehicles, spaceflight expertise, and orbital hardware would become available to the international space community for commercial satellite launches.

The first component of ISS, the Russian-built control module Zarya, was sent into orbit by a Proton rocket launched from Baikonur in Kazakhstan, on November 20, 1998. Then came the U.S.-built Unity node, and the Russian Zvezda service module, just the first three of a long series of components. What began as a pragmatic foreign policy measure has since yielded valuable benefits, not in space but on the ground. Russia and the United States have what might be described as a difficult relationship. Orbital cooperation helps maintain at least a few friendly channels of communication. The loss of Space Shuttle *Columbia* and its crew in February 2003, and the backup of Russia's reliable workhorse, highlighted the value of international cooperation in times of crisis.

These geopolitical factors go a long way toward justifying the entire ISS. They strengthen friendships and alliances, contribute to the global economy, and enhance the prospects for long-term peace. But the tremendous societal benefits of ISS are only one aspect of its character. Ultimately, it is a spacecraft with a job to do. Its primary mission is to expand scientific knowledge that can be translated into innovative applications for the development of new materials, new therapies in medicine, and in the preparation for future human missions into deep space. Whether or not these missions will continue to be truly international remains to be seen.

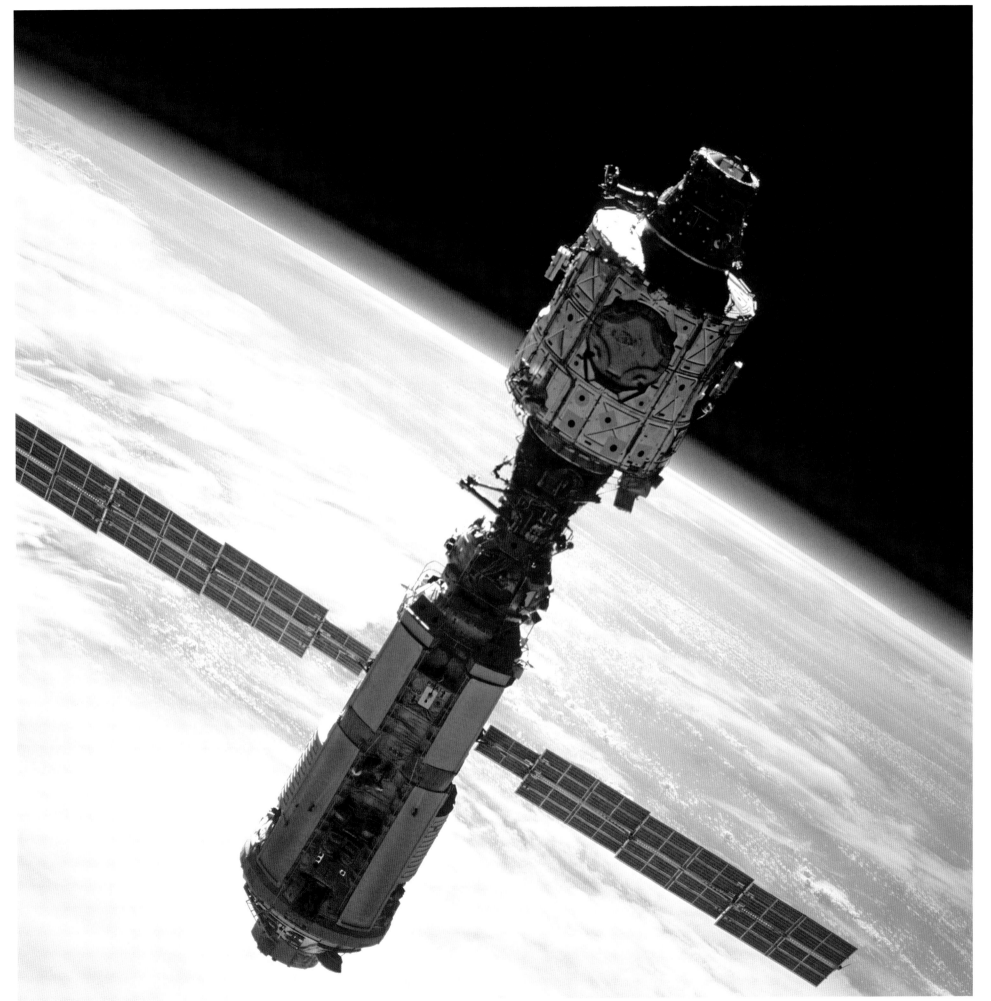

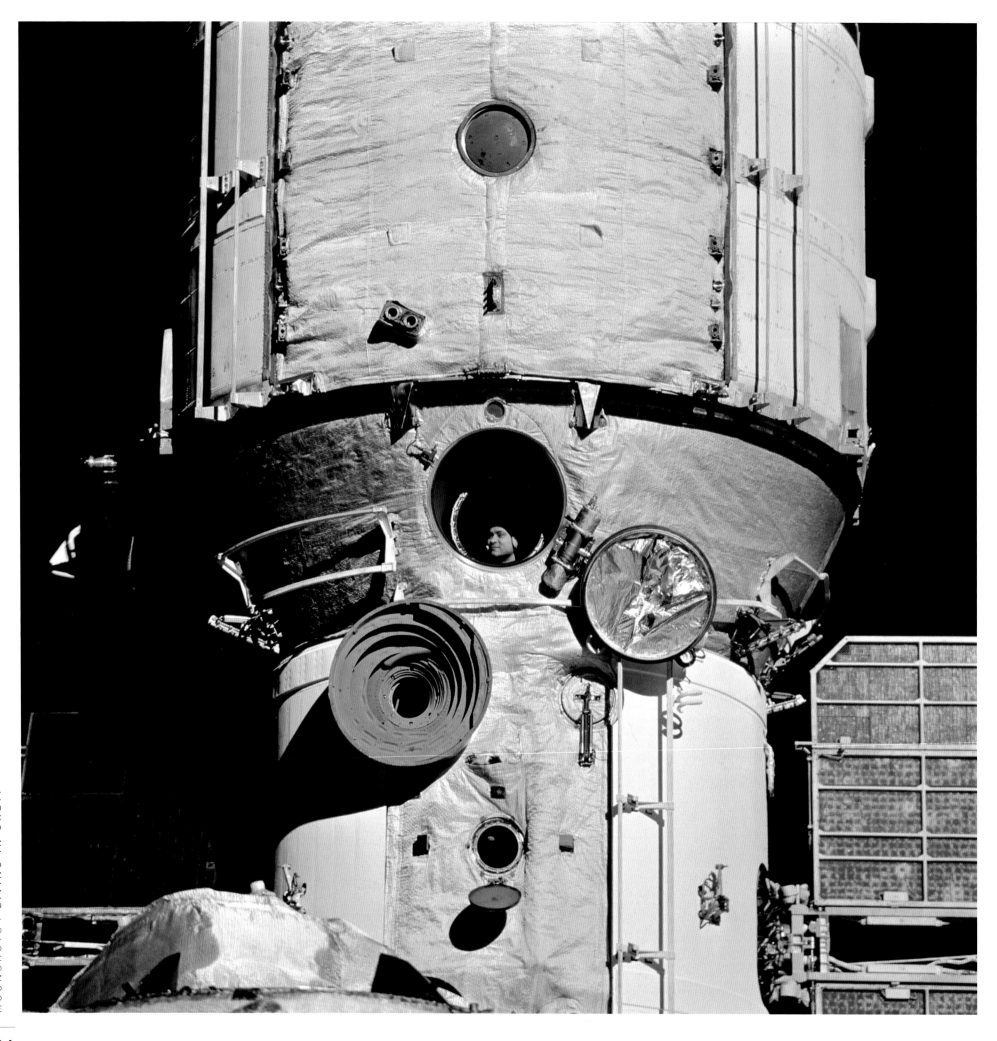

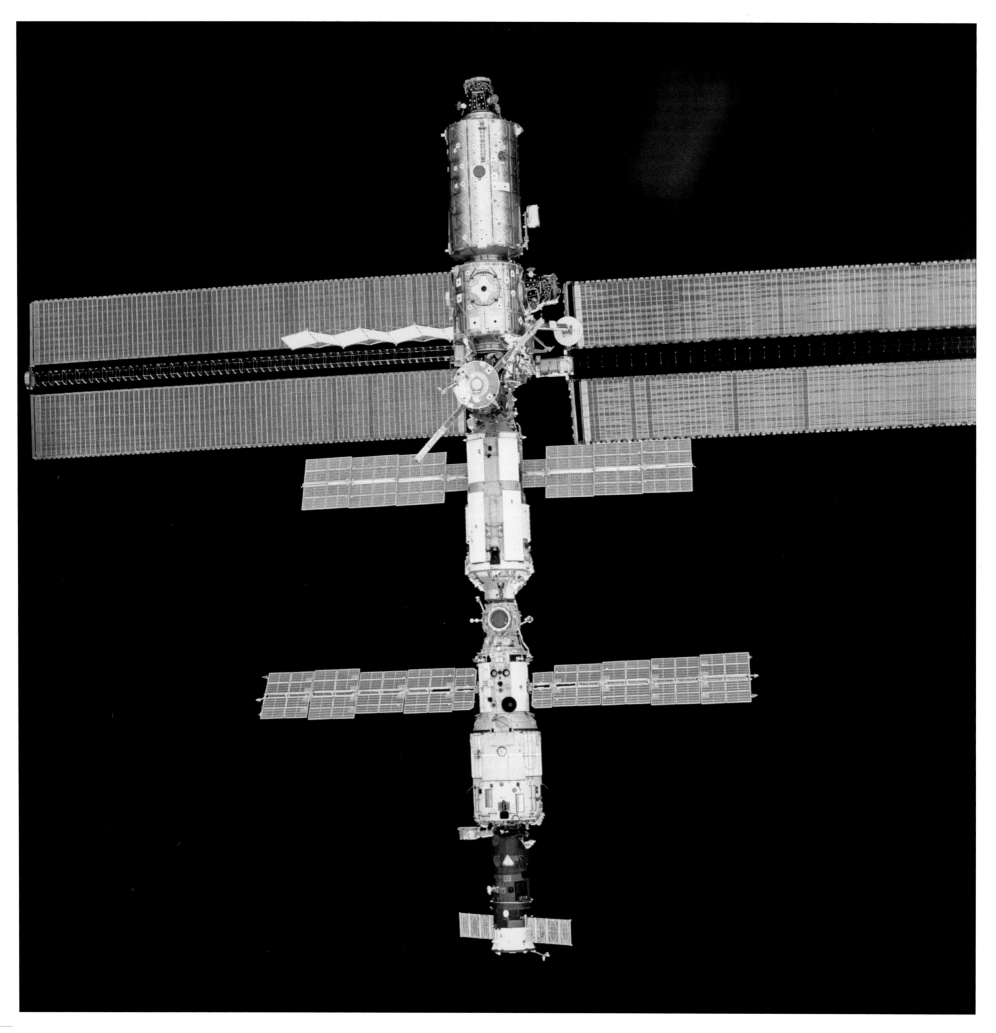

INDEX

Page numbers in italics indicate an item in a photograph or caption.

Aaron, John, 128
Advanced Research Projects Agency (ARPA), 17
Agena target vehicle, 24, *39–41*
Akers, Thomas, 229, *239*
Alcock, John, 8
Aldrin, Buzz, 8, 10, *47*, *49*, 94–95, *108–109*, *112–119*, 128
Allen, Joe, 215
Anders, Bill, 78
Anderson, Michael, 230
Ansco Autoset 35-millimeter, 22
Antares landing module, *146–148*, 147, *150–151*
Apollo Project, 8, 18, 20, 94–97
Apollo 1, 52, *60–61*
Apollo 7, 19, 53, *54–59*
Apollo 8, 19, *50*, 78, *82–85*
Apollo 9, 19, *63–75*
Apollo 10, *86–88*
Apollo 11, 8, 19–20, *97–99*, 100, *103–108*
Apollo 12, 21, 128
Apollo 13, 20, 139–140, *142–145*
Apollo 14, 146–147, *148–153*
Apollo 15, 156, *157–167*
Apollo 16, 168, *169–181*, 170–172
Apollo 17, *182–183*, 183–184, *185*, *186–191*, *193–209*
Apollo command module (CM), 53, *86–87*, *102–103*, *124*, *144*
Apollo Guidance Computer (AGC), 21, 95, 128
Apollo lunar surface experiment package (ALSEP), 170
Apollo-Soyuz docking mission, 21
"Big Red Booster", 77
lunar module (LM), 19, 53, *58*, *63–64*, *70–73*, 78, *88*, *103*, *106*, *109–111*, *113*, *116–119*, *125*, *130–131*, *142–143*, *146–148*, *150–151*, *185*, *207*

Lunar Roving Vehicle, *155*, *157–158*, *161*, *165*, *170–175*, *178*, *190*, *196–199*
modular equipment transporter (MET), 147, *152*
scientific instrument module (SIM) bay, 156, 172, *206*
Apollo Soyuz Test Project (ASTP), 223, *224*, 225
Aquarius landing module, 140, *142–143*
Armstrong, Neil, 8, 10, 24–25, *41*, *48*, 94–95, *108–109*, *112–120*, 121, 128

Bales, Steve, 94
Bean, Al, 128, *131–134*
Borman, Frank, 34, 53, 78
Brand, Vance, 215, 225
Brown, Arthur, 8
Bush, George W., 230

cameras, 22, 184
Carr, Jerry, 212–213, *215*
Casper command module, 168, 172
Cernan, Gene, 25, *42*, *46*, 79, 183, *186–190*, *196–197*, *205*, *207*
Chaffee, Roger, 18–19, 52, *60–61*
Charlie Brown command module, 79
Chawla, Kalpana, 230
Clark, Laurel, 230
Clarke, Arthur C., 21, 97
Cold War, 16–17, 21
Collins, Michael, 10, 19–20, 94, 107
Columbia Accident Investigation Board (CAIB), 230
Columbia command module, 94–95, *102–103*, *124*
Conrad, Pete, 128, *133*, 212
Cuban missle crisis, 17–18
Cunningham, Walter, 19, 53, *54–55*, 59

digital scanning, 12
Duke, Charlie, 94–95, 168, 171–172, *176–177*, 178

Eagle landing module, 94–95, *106*, *109–111*, *113*, *116–119*, *125*
Earth rises, *82–83*
Eisele, Donn, 19, 53, *54*
Endeavour command module, 156, *164–165*
Evans, Ron, 183

Falcon landing module, 156, *157*
Feynman, Richard, 228–229
film, 22, 184
Free Return Trajectory, 140
Full Moon (Light), 12

Gagarin, Yuri Alekseyevich, 17
Gardner, Dale, *238*
Garriott, Owen, *220*
Gemini program, 20, 24–25, *27–33*
Gemini III, 24, *28–29*
Gemini IV, 24, *27*, *30–33*
Gemini IX, 25, *42–43*, *46*
Gemini VI, 34, *34–38*
Gemini VII, 34, *34–38*
Gemini VIII, 24, *39–41*, *48*
Gemini XI, *45*
Gemini XII, *44*, *47*, *49*
Gibson, Ed, 212
Glenn, John, 22
Goddard, Robert, 16
Gordon, Dick, 128
Grissom, Virgil "Gus", 18–19, 24, *28–29*, 52, *60–61*
Gumdrop command module, 78

Haise, Fred, 139–140, *141*
Hasselblad 500EL, 22
Hasselblad 550C, 11, 22
Hasselblad EL, 22
Herrington, John, *253*
Hoffman, Jeffrey, 229
Hubble space telescope, 229–230, *242–243*
Husband, Rick, 230

International Space Station, 21, 184, 186, 213, 215, 230, *239*, *244*, 245–246, *247–253*
   Destiny science module, *252*
   Zarya control module, 246
   Zvezda service module, 246
Intrepid landing module, 128, *130–131*
Irwin, James, 156, *157*

Jarvis, Greg, 228
Johnson, Lyndon, 18
Johnson Space Flight Center (Houston, TX), 10

Kennedy, John F., 16–18, 223
Kerwin, Joe, 212
Kitty Hawk command module, 146
Komarov, Vladimir, 53–54
Korolev, Sergei Pavlovich, 53–54
Kosygin, Alexei, 223
Kranz, Gene, 94–95
Kubasov, Valeri, 225

Leonov, Alexei, *224*, 225
Life magazine, 10, 17
Light, Michael, 12
Lindbergh, Charles, 8
Logsdon, John, 17
Lopez-Alegria, Michael, *253*
Lousma, Jack, *221*
Lovell, Jim, 10, 34, 78, 139–140, *141*
lunar module (LM), 19, 53, *58*, *63–64*, *70–73*, 94–95

Mailer, Norman, 20, 96, 184
Manned Maneuvering Unit (MMU), *236–239*
Mars rovers, 21
Mattingly, Ken, 168, 172
McAuliffe, Christa, 228
McCandless, Bruce, *236–238*
McCool, William, 230
McDivitt, Jim, 24, *76*, 78
McNair, Ron, 228
Mercury spacecraft, 17, 18, 22

Mir complex, 245, *248–250*
Mitchell, Ed, 146–147, 151, *151*
moon
   Apennine Mountains, 156, *159–160*
   Dead Man's Zone, 94
   Fra Mauro, 140, 147
   Genesis Rock, 156
   Hadley Rille, 156, *162*, *166–167*
   Mare Imbrium, 156
   Oceanus Procellarum, 128
   Shorty Crater, 184
   Spur Crater, 156
   surface of, *84–85*, *89–91*, *122–123*, *129*, *135*, *153*
   Surveyor crater, 128
moon landing, 19, 94–97, *110–125*
Musgrave, Story, 229

NASA, 18
   Apollo 1 fire, 52–53
   Apollo program, 18–19
   Challenger explosion, 54, 228
   Columbia disaster, 54, 230
   cooperation with Russia space program, 223, 225
   lunar module (LM), 19
   moon landing, 19
   moon landing in popular memory, 95–97
   Orion spacecraft, 21
   public interest and, 184, 186
   1947 Soyuz disaster and, 53–54
   Webb and, 18
   See also Apollo Project; Gemini program
National Geographic, 10
Nikolayev, Andrian, 34
Nixon, Richard, 223, 226

Obama, Barack, 230
Oberth, Hermann, 16
Odyssey command module, 140, *144*
Onizuka, Ellison, 228
Orion landing module, 168, 170–172, *174*, *179*

Orion spacecraft, 21, 230

Pogue, Bill, 212, 213, 215, *215*
Polyakof, Valery, *250*
Popovich, Pavel, 34
Portable Life Support System (PLSS) backpack, *69*
Powers, John "Shorty", 17

Ramon, Ilan, 230
Reagan, Ronald, 228, 245
Resnik, Judy, 228
Ride, Sally, *234*
Roosa, Stuart, 146–147
Russian space program, 230

Saturn 1-B rocket, *80*, 212, 215, *222*
Saturn V, 19, *50*, 78, *81*, *105*, *126–127*, 128, *137*, 212, *213*, 215, 226
Schirra, Wally, *9*, 11, 19, 22, 34, 53, 56
Schmitt, Harrison, *182–183*, 183, *191*, *196–197*, 200, *201–202*, 204
Schweickart, Rusty, *69*, 78
Scobee, Dick, 228
Scott, Dave, 24, *48*, *65–68*, *76*, 78–79, 156, *158*, *161*
Shepard, Alan, 18, 146–147, *149*, *152*
Sidey, Hugh, 17
Skylab Orbital Workshop, *210–211*, 212–213, *214–221*, 215
Slayton, Deke, 215, *224*, 225
Smith, Mike, 228
Snoopy landing module, 79
solar eclipse, *136*
Soviet space program, 77–78, 223, 225
   Mir complex, 245, *248–250*
   N-1 rocket, 77
   Proton rocket, 77, 246
   Salyut orbiting platforms, 245
   Soyuz capsules, 223, 225
   Zond 5, 77–78
Soyuz project, 230
   1947 Soyuz disaster, 53–54

Space Race, 16–17, 77–78, 223, 245
Space Shuttle program, 8, 226, 228–230, *231–244*
   Atlantis, 228, *251*
   Challenger, 54, 228–229
   Columbia, 54, *227*, 228, 230, 246
   Columbia Accident Investigation Board (CAIB), 230
   Discovery, 228–229
   Endeavour, 229–230
   Rogers Commission, 228–229
Spider luner module, 78
Sputnik satellite, 16
Stafford, Jim, 34
Stafford, Tom, 25, *42–43*, 79, 215, 225
Surveyor 3, *134*
Swigert, Jack, 139–140, *141*, *143*

Thornton, Kathryn, 229, *239*
Tsiolkovsky, Konstantin, 16

Von Braun, Werner, 16
Vostok capsule, 16–17, 18

Webb, James Edwin, 18, 52–53, 77
Weinberger, Caspar, 228
Weitz, Paul, 212
White, Ed, 18–19, 24, *30–33*, 52, *60–61*
Wiesner, Jerome, 17
Worden, Al, 156, *164*

Yankee Clipper command module, 128
Young, John, 24, *28–29*, 79, 168, *169*, 170–171, *170–175*, 178

Zond 5, 77–78

CREDITS

This book was made possible by the collaboration of NASA, and its generous provision of images. NASA staffers Steven Dick, Colin Fries, Mike Gentry, Connie Moore, and Margaret Persinger were endlessly patient with my requests for information about archival sources. A number of wonderful image historians and archivists elsewhere were also incredibly generous with their time and resources. I would like to thank Kipp Teague, Paul Fjeld, Ed Hengeveld, Eric Jones, Mike Marcucci, Jack Pickering, and Mike Constantine (whose book *Moonpans* is essential for any Apollo enthusiast). I also acknowledge the archival and digital scanning work of the School of Earth & Space Exploration at Arizona State University.

Thanks also to the Hasselblad camera company in Sweden for its friendship and support. It should be noted that *Moonshots* is an independent production and not in any way an official commercial or editorial product of Hasselblad. Even so, the assistance of the company's PR department, and its goodwill toward this project, was welcome. Thanks in particular to Ida Gustafsson and Hasselblad Foundation librarian Elsa Modin. Additionally, I would like to acknowledge the generous support of the Smithsonian Institution's National Air and Space Museum in Washington, D.C., for superb close-up images of Gemini and Apollo hardware on pages 13, 14, 27, 94, and 95.